AF707922

Biocontrol of Plant Disease

SCIENCES

Ecosystems and Environment, Field Director –
Françoise Gaill and Dominique Joly

Environment, Natural or Anthropogenic Pressures,
Subject Head – Denis Faure

Biocontrol of Plant Disease

Recent Advances and Prospects in Plant Protection

Coordinated by
Claire Prigent-Combaret
Bernard Dumas

ISTE

WILEY

First published 2022 in Great Britain and the United States by ISTE Ltd and John Wiley & Sons, Inc.

Apart from any fair dealing for the purposes of research or private study, or criticism or review, as permitted under the Copyright, Designs and Patents Act 1988, this publication may only be reproduced, stored or transmitted, in any form or by any means, with the prior permission in writing of the publishers, or in the case of reprographic reproduction in accordance with the terms and licenses issued by the CLA. Enquiries concerning reproduction outside these terms should be sent to the publishers at the undermentioned address:

ISTE Ltd
27-37 St George's Road
London SW19 4EU
UK

www.iste.co.uk

John Wiley & Sons, Inc.
111 River Street
Hoboken, NJ 07030
USA

www.wiley.com

© ISTE Ltd 2022
The rights of Claire Prigent-Combaret and Bernard Dumas to be identified as the authors of this work have been asserted by them in accordance with the Copyright, Designs and Patents Act 1988.

Any opinions, findings, and conclusions or recommendations expressed in this material are those of the author(s), contributor(s) or editor(s) and do not necessarily reflect the views of ISTE Group.

Library of Congress Control Number: 2022941466

British Library Cataloguing-in-Publication Data
A CIP record for this book is available from the British Library
ISBN 978-1-78945-098-9

ERC code:
LS6 Immunity and Infection
LS6_8 Infectious diseases in animals and plants

Contents

Introduction

Promoting Biocontrol and Sustainable Crop Protection Strategies, a Major Challenge for the 21st Century

Bernard DUMAS[1] and Claire PRIGENT-COMBARET[2]

[1] *LRSV, Toulouse University, France*

[2] *EM, UMR 5557, Lyon 1 University, France*

During the 20th century, in response to the need to significantly increase production, agriculture resorted to the massive use of chemical inputs (fertilizers and pesticides). This made it possible to provide crops with all the essential nutrients needed for their growth (nitrogen fertilizers, phosphorus, etc.) while effectively protecting them from diseases and pests (fungicides, insecticides, herbicides). Although these products have been extremely effective and have resulted in a significant increase in crop yield, their large-scale use has led, in some cases, to the degradation of soil quality and has had a dramatic impact on natural flora and fauna (a reduction in biodiversity, the appearance of resistant individuals, etc.). Today, the increased availability of better-quality food has made it possible to achieve food safety and has also provided customers with wide access to a healthy diverse diet. These beneficial effects have, however, come at the cost of adverse health effects associated with the use of certain inputs with a hazardous toxicological profile. As a result, the way in which we develop and use agricultural inputs has seen a major reorientation following a shift in policies to drastically reduce agricultural inputs originating from synthetic

coordinated by Claire PRIGENT-COMBARET and Bernard DUMAS. © ISTE Ltd 2022.

chemistry (including chemical fertilizers and phytosanitary products), which we have been systematically implementing for the last several years to focus on developing alternative solutions with a lower environmental impact. The main goal of these policies is the sustainable production of quality foods that are safe for both the environment and the consumer.

Twenty-first century agriculture must therefore face the added constraint of reducing environmental impacts in addition to ensuring adequate production capacity to maintain the viability of its economic model. There are enormous economic challenges to be addressed for developing a sustainable agriculture that also respects the environment. Today, consumers are increasingly aware of the need to produce healthy food albeit with a low environmental impact. To this end, they are increasingly turning to products from alternative systems; these may be products with zero residue specifications from organic farming, or local production sold via short supply chains. This calls into question our historical production model with respect to its choice of crop species, inputs used and marketing channels.

To address these new challenges, a significant research and development effort is needed to optimize and integrate new agronomic methods so that we can circumvent the massive use of non-natural products originating from so-called "conventional" synthetic chemistry. Such research involves setting up new farming practices (tillage, crop rotation, etc.) and the genetic selection of varieties that guarantee yield and improve resistance to stress and which are in symbiosis with soil microbial communities, while at the same time developing new inputs with a low environmental impact. In this context, the implementation of effective solutions as an alternative to "conventional" treatments will be decisive for the competitiveness of our current agricultural model.

One approach that is currently being developed is through the use of natural compounds to combat weeds, diseases and pathogens (biocontrol) and to optimize nutrition and plant development (biostimulant). For several years now, the application of these two types of products has seen a sharp increase, particularly in the context of organic farming. Manufacturers in this sector have set themselves the ambitious goal of capturing 30% of the crop protection market by 2030, which is in line with national public policies, notably in the European Union (using the strategy "Farm to Fork", which aims for a 50% reduction in plant protection products and 25% conversion to organic farming by 2030 – https://ec.europa.eu/food/farm2fork_en). The major obstacle to achieving these ambitious objectives, however, remains identifying new active substances or living organisms of agronomic interest which are more environmentally friendly. Moreover, in order to exploit the potential of these active substances and provide solutions to ensure optimal crop protection while guaranteeing better yield, we also need to understand their mode of action in the complex environment that constitutes our agricultural system (Figure I.1).

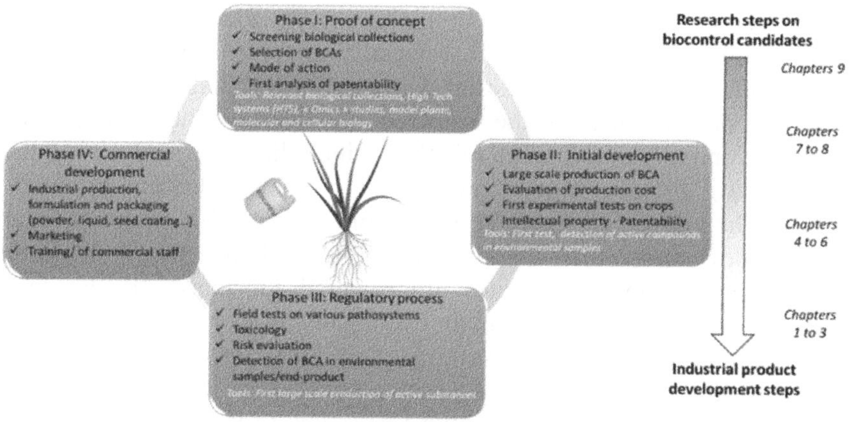

Figure I.1. *Necessary steps in the development of biocontrol agents (BCAs)*

The objective of this book is to gather works written by leading scientists in their fields in the form of chapters to illustrate the multifaceted aspects of the research devoted to finding new environmentally compatible solutions to protect plants against diseases, while maintaining crop yields. This book also addresses the important question of the current regulatory process needed to launch plant production products on the market (Figure I.1). We have chosen articles from research works presenting new advances on plant disease management through innovative strategies. Since an exhaustive panorama of biocontrol strategies is out of the scope of this book, certain topics such as the mechanisms involved in the protection of plants against insects by indirect action (e.g. the use of pheromones and kairomones and other natural defense stimulators) will not be covered in this book. However, the ability of biocontrol agents to protect plants against bacterial, fungal or oomycete pathogens or diseases triggered by insects or nematodes directly (e.g. by producing antimicrobial peptides, using the quorum quenching strategy with microorganisms, using plant or agro-industrial by-products as biopesticides, etc.) or indirectly (e.g. by increasing plant defense signaling pathways (via induced systemic resistance (ISR)), or by stimulating plant growth and development) will be described. We also want to address new strategies such as the development of phage-based biocontrol and those in which preventing pathogen-induced dysbiosis of plant microbiota is considered to be key to ensuring the overall health of the plant.

The different phases in the development of BCAs are identical to those when developing chemicals. The first phase is to identify candidates by screening biological collections that usually involve collaborative work between academic labs and industrial partners. At this stage, it is crucial to analyze the patentability of the selected candidates. The second phase concerns the initial development of BCAs, including the analysis of their activity on target crops and the evaluation of their industrial production feasibility. The knowledge acquired during phase I and II is essential for the subsequent development of the products. Phases I and II are intimately linked and only a few candidates are selected for phase III. Phase III (regulatory process), done on the main candidates, is probably the one that is the most expensive and time consuming. This phase involves toxicological analyses and several years of field tests to demonstrate the efficacy of the product. Finally, phase IV includes the final steps to launch the product on the market. Together, these phases need about 10 years and an investment of several hundreds of millions of euros to be completed. Even if the investment is mainly in phase III, regulations defined by authorities influence the decision to select a certain BCA (phase I) and to continue with its initial development (phase II). The positioning of the chapters in relation to these different stages is indicated on the right side of the figure.

A crucial step in introducing a plant protection product onto the market is the associated regulatory process, which directly impacts the investment needed to launch the product. The way in which this regulatory process is defined also influences the search for new active compounds in that those that have the greatest chance to be homologated will be preferred (Figure I.1). For these reasons, the first part of this book focuses on issues related to commercial biocontrol compounds. First, a description of the rules and regulations for the commercialization of biocontrol products is given by Robin et al. (Chapter 1). This chapter starts by giving definitions of the term "biocontrol" and also of other denominations used in the context of plant protection products (PPP) as required by the European regulation (e.g. "BioControl Agents", "Active Substances", "Biorationals", etc.) even though the word "biocontrol" is not used as such. This chapter also discusses the problematic issues in these regulations regarding the use of biocontrol substances and the slow and long path to gaining the approval of biocontrol products in the PPP regulation. Limitations in the French and European regulations on so-called "biocontrol" products and the different possible suggestions to reform these regulations are also discussed.

Chapter 2 by Guibert et al. gives an overview of the various biocontrol products used in horticulture. Horticulture is a vast agronomic sector involving the cultivation of fruits, vegetables and ornamental plants. It is a major market for biocontrol products and targets both fresh and, in some cases, perishable products to improve shelf life. In the case of edible products, plant protection strategies should scrupulously avoid contamination by toxic substances and efforts should also be dedicated to preserving the gustative quality of the food in addition to conserving their nutritional value. In this sector, consumers are strongly concerned about the safety aspects of food products, as confirmed by the continuously increasing demand for organic farm produce. Although biocontrol has been integrated into horticultural practices for more than 40 years, the reduced or banned application of phytopharmaceutical products (e.g. glyphosate) has motivated cultivators to make fundamental changes and has also challenged many entrenched agricultural practices. The increase in the use of biocontrol methods and biostimulants, which is now under way, is largely driven by the greater awareness in society of the negative impact of pesticides, as described in Chapter 2. The authors describe the different trade-offs that must be considered in horticulture and the difficulties that farmers could encounter: the trade-off between plant growth and plant protection against disease as well as economic trade-offs linked to eco-innovations (increased costs, productivity gains, critical importance of certification and labeling on the marketplace, etc.). Further studies on the safety and environmental sustainability of biocontrol products are still needed before their deployment on the large scale as an alternative to PPP.

Chapter 3 by Ghosson et al. discusses this issue in detail. While biocontrol products are generally considered to be safer for the environment than chemical products, they are, by definition, biologically active products, and it is therefore necessary to analyze their environmental fate following their deployment. This aspect is related to regulatory rules that apply to biocontrol products, which are largely inspired by the existing rules for chemicals. However, a natural product is much more complex than its chemical alternative which is generally composed of a nearly pure molecule with formulants and adjuvant, and whose degradation products can be traced. Plant and microbial extracts contain thousands of metabolites, and it is almost impossible to predict the diversity of compounds produced by a living microorganism. As noticed by Ghosson et al. a way to circumvent these difficulties could be to apply "omics" or "meta-omics" strategies, notably genomics and metabolomics, to get a wider overview of a given product in the "omic" environment.

The second part of this book focuses on the development of biocontrol products based on active natural compounds obtained from plants (also known as botanicals) and microorganisms. Chapter 4 by Ntalli and Caboni presents a survey of the literature from the last 10 years on plant-based products whose ability to control root-knot nematodes has been assessed in vitro or in vivo, including in fields. Nematodes belonging to the *Meloidogyne* genus are major agricultural pests, and developing effective biocontrol strategies to combat root-knot nematodes is currently the subject of much research; significant advances have been made in recent times. Nevertheless, molecular targets within the nematode on which botanicals interact have only rarely been identified. In this context, Ntalli and Caboni describe few recent works which have investigated the mode of action of plant secondary metabolites on root-knot nematodes.

Agro-industrial wastes and by-products are inexpensive and abundant sources of bioactive molecules, notably those from the agroforestry sector or other industries that process plant materials. These products contain antioxidants, antimicrobials, insecticides and nematicides. With a special emphasis on nematicide activities, Chapter 5 by Andres and González Coloma presents the activities of agro-industrial wastes, which have been the subject of intense study in recent times. Such wastes include biochar produced from the pyrolysis of wood and other plant materials, by-products from the production of essential oils (hydrolates or hydrosols), and wastes from olive oil production. Through these examples, Andres and González Coloma suggest that agro-industrial wastes constitute an almost inexhaustible source of biopesticides, but more research is needed to improve their efficiency and reduce the potential phytotoxicity of these products.

We end the second part with Chapter 6 by Montesinos et al. which focuses on the development of novel plant disease protection strategies based on the use of antimicrobial peptides and defense elicitor peptides. Plants, animals and microorganisms synthesize a wide diversity of peptides that have antimicrobial properties, or are able to trigger the immune response of the plant. Several peptides have already shown their activity against plant pathogens in vitro or in planta under greenhouse conditions, but reports on their efficiency in fields are still scarce. Such peptides could be produced naturally or synthetically via heterologous expression systems implemented in plants or microorganisms. However, there are several challenges, including their facile synthesis in a cost-effective manner, improving their delivery/formulation and stability, all of which determine their efficiency in plant disease protection. A clear regulatory framework for their application as biocontrol products is also needed.

The third and last part of this book addresses the question of how biocontrol products may impact pathogen microbial development, notably by exploiting microbial signaling and microbe–microbe interactions. Biocontrol of plant pathogens via quorum quenching is discussed in Chapter 7 by Faure and Latour. Quorum quenching (QQ) refers to any of the mechanisms involved in the degradation of quorum-sensing (QS) molecules, whose concentrations increase with the proliferation of QS-emitting populations and thus act as signals. These signal molecules, when reaching a concentration threshold, can bind and activate transcriptional regulators to control the expression of QS-regulated genes, especially those involved in the biosynthesis of virulence factors, in several bacterial pathogens. Quorum quenchers can alter either of the two main steps in QS signaling pathways, namely the QS synthesis or the accumulation and interaction of QS signals with their regulators. Accordingly, there are two main types of quorum quenchers: QS inhibitors and QS-degrading enzymes having QQ activities and bacterial populations sharing this property that could be used as biocontrol agents. These agents target virulence factors of plant pathogens, but do not inhibit their population growth. QQ activity can be stimulated by using treatments that stimulate the proliferation of externally introduced or native QQ populations. Thus, QQ is an emerging novel biocontrol trait that is widely distributed in microbial soil communities. The efficiency of QQ treatments in combination with antibiosis approaches is currently under evaluation in field assays, and promising results are expected from these new biocontrol strategies.

Another strategy to counteract pathogenic development is by using bacteriophages. Bacteriophages (or phages) are an essential and often underestimated components of plant microbiota, notably in the rhizosphere. Due to their pathogenicity towards phytopathogenic bacteria and their specificity, phages are potentially attractive biocontrol agents. As described in Chapter 8 by Clavijo-Coppens et al. a number of phages targeting major phytopathogenic bacteria belonging to various genera (*Acidovorax, Burkholderia, Erwinia*, etc.) have been characterized and some of them have already been shown, at least in experiments under controlled conditions, to successfully control bacterial diseases on agronomically relevant plants. However, these encouraging results should be taken with caution in the face of technical challenges, which could hamper the use of phage preparations in agriculture. The cost of large-scale preparation of phage-based products has to be compatible with the overall cost of the commercial end-product in order to be acceptable to farmers, and progress also needs to be made in the formulation of phage-based products, notably regarding phage viability. An attractive approach, as for other biocontrol substances, could be coating seeds with a phage preparation, which could be an elegant way to protect plants while avoiding subsequent treatments in the field. Finally, the authors discuss the regulatory challenges faced by industrial actors who are willing to commercialize these new products. Despite the great potential of this strategy, there is little doubt that a deeper understanding of the role of phages in the plant ecosystem will be necessary to encourage the development of phage-based products.

Up to now, the development of biocontrol products has been mainly viewed as a replacement of chemicals by natural products exhibiting antimicrobial activities, hoping that these compounds are less harmful to the environment because of their natural origin. However, plant diseases can also be tackled by optimizing the functioning of the plant microbiota, which increasingly is being recognized as an essential component of plant health. Plant phyllosphere and rhizosphere host a myriad of microorganisms collectively designated as the microbiota. The microbiota is essential for plant life and fitness: it is a key contributor to plant nutrition and its resistance to biotic and abiotic stresses. In the last decade, metabarcoding investigations have pointed out the diversity of plant microbiota (rhizosphere, phyllosphere, spermosphere) in various plant species growing in contrasted environmental conditions. Recent results have shown that the composition of microbiota is in part, determined by the plant, notably with respect to the composition of plant exudates, and can be tuned to respond to new environmental conditions, and that certain microbial species can play a tremendous role in altering the biological function of the microbiota. In agricultural systems, chemical inputs which have been massively used for close to a century, has caused soil depletion and has had a dramatic impact on the diversity and functioning of plant microbiota. Thus, a major challenge of modern agriculture is to reconcile the highly reduced use of chemicals while maintaining high yields in agricultural environments that is today also strongly impacted by urbanization and climate change. One solution could come from the use of microbial formulations to help plants get nutrients and combat diseases while allowing the soil to recover its properties. However, the use of active microbial strains in agricultural systems has thus far not been optimized. Chapter 9 by Yu et al. addresses this essential issue and, on the basis of the latest progress made in this field, shows that new strategies could be proposed in the near future to optimize plant microbiota.

To conclude, these contributions together offer us a large panorama of the various strategies that are currently available or being developed to offer alternative environmentally safe products to combat plant diseases. They also point out the bottlenecks that hamper the introduction of these compounds into the market, notably the lengthy regulatory process that, at least in European countries, does not distinguish natural and chemical pesticides. However, the urgent need for alternative solutions to replace conventional products will undoubtedly support strong investment in this research field. Nevertheless, the road ahead is still long in facilitating regular use and improving the efficiency of biocontrol strategies. The transition from results acquired under greenhouse conditions to those acquired in fields over several consecutive years is still the most difficult step in the development of a biocontrol product. However, high magnitude and quality of research efforts on biocontrol strategies worldwide allows us to be optimistic about being able to provide farmers with novel nature-inspired biomimetics and efficient biocontrol solutions.

1

Regulatory Aspects of Biocontrol

Diane ROBIN, Léa MERLET and Patrice MARCHAND
ITAB, France

Within the term "biocontrol", only "biological control" thus macroorganisms (typical concept), "BioControl Agents" (BCAs) encompassing the three pillars outlined by Plant Protection Products Regulation EC 1107/2009 (namely microorganisms, semiochemicals and natural substances of plant, animal, mineral and microbial origin) and macroorganisms (fourth pillar) are considered. Macroorganisms are currently unregulated, except in France. For those considered as Plant Protection active substances, an approval pathway through EU Regulations follows general rules although some of these may be waived, depending on the pillar. In fact, 216 BioControl active substances are approved at the EU level, representing around 48% of total active substances (38% in 2011). Globally, and in a fairly stable way since 2011, around 50% are natural substances, 1/3 are microorganisms and 20% are semiochemicals. However, this significant progression, taking into account that 19 BioControl active substances were removed during the same period of time, is hiding the fact that many BioControl candidate substances are not being approved due to the inability to evaluate these substances (especially natural substances), which has left dozens of applicant substances unapproved.

Biocontrol of Plant Disease,
coordinated by Claire PRIGENT-COMBARET and Bernard DUMAS. © ISTE Ltd 2022.

1.1. Regulatory definition of biocontrol

1.1.1. *Definitions of biocontrol*

Biocontrol can be considered as including only biological control and macroorganisms, which follows the typical concept in the English-speaking world, or alternatively all the BioControl Agents (BCAs) encompassing the three pillars: microorganisms, semiochemicals and natural substances (of plant, animal, mineral and microbial origin) and occasionally including macroorganisms (fourth pillar), following the rather French and European concept set out by the International Biocontrol Manufacturers' Association (IBMA) (Robin and Marchand 2019a). Sometimes this definition is extended to physical barriers (nets, monitoring traps, etc.), but this thinking has no specific regulations to date.

1.1.2. *Applicable regulations*

Biocontrol, within its BCA limits (BioControl Agents and Biocontrol products), has been under the control of the phytopharmaceutical regulation EC 1107/2009 (EC 2009b) at the European level, since 2011. Macroorganisms are not regulated in EU but when species are not endogenous, they are regulated in France (Robin and Marchand 2020).

1.1.2.1. *Denominations – wording*

Therefore, these products, and the substances which support them, even if they are not defined in such a way, are considered as "pesticides" sometimes with the alternate qualifiers of "biological pesticides", "biopesticides", "biorationals" or "BioControl agents" (BCA) in English.

1.1.2.2. *Denominations – glossary*

"Active substance" (a.s.) – substance from the three pillars regulated by Plant Protection Products Regulation EC 1107/2009 and listed in Regulation EC 540/2011.

"Product" – formulated approved active substance with market authorizations.

"Biocontrol product" – product with biocontrol active substance from the three pillars.

"Semiochemicals" – semiochemicals are defined by the European commission as substances or mixtures of substances emitted by plants, animals and other organisms that evoke a behavioral or physiological response in individuals of the same (= pheromones) or other (= allelochemicals) species. Natural-identical synthesized molecules are also included in this definition.

"Natural substances" – active substance of plant, mineral, animal or microbial origin, non-transformed or activated.

"Biological control" – generally dedicated to macroorganism uses.

"Biorationals" – generally considered as "natural substances" or of biological origin.

"Biocontrol" – may also be attributed to crop protection by living organisms: macroorganisms and microorganisms.

"BioControl Agents" or "BCAs" – commonly attributed to all biocontrol plant protection active substances.

1.2. Current issues and limitations

First, it has to be mentioned that the word "biocontrol" is absent in the Plant protection Products (PPP) regulation (EC 2009a). The notion of biocontrol is therefore external to the regulation that manages ¾ of the categories. Moreover, the word "biocontrol" is also absent in the Sustainable Use of pesticides Directive (SUD) (EC 2009b), although integrated pest management is mainstream, and biocontrol is one form of this concept. Consecutively, the concept of biocontrol or the biocontrol qualification of the substances and the corresponding product is from an external point of view; consequently, there is no official specific pathway or wavers during approval or renewal.

1.2.1. *The 3 PPP pillars*

During our study on their evolution, we showed at the ITAB institute that "biocontrol" active substances (a.s.) can belong to all parts from A to E (Robin and Marchand 2019b), and all types (active substance, low-risk, basic or candidate for substitution) of PPP regulation (Robin and Marchand

2021a). The notion of biocontrol applied to these substances is therefore diluted, and these substances diluted in EC Regulation 540/2011 and its sub-parts without distinction (Robin and Marchand 2021b) although the concept and the word "biocontrol" was not written in the PPP regulation. Many of these natural substances with non-biocidal properties and modes of action (MoA) are of very low toxicological concern and therefore do not have maximum residue limits (MRLs) (Charon et al. 2019). However, even if some would want to generalize that belief, their toxicity is not always trivial.

1.2.2. *Fourth pillar*

Global biocontrol includes macroorganisms, which fall under neither PPP regulation nor any specific regulation at the European level. At the national level, only France regulates transfers of non-native macroorganisms from mainland France to overseas territories, including Corsica (Journal Officiel "Lois et Décrets" (JORF) 2022), and vice versa (Robin and Marchand 2020). It aims to avoid the potential problems of the propagation of non-native macroorganisms, which could become invasive species outside their indigenous territory. Use is granted when this risk is guaranteed to be null (inadequate adaptation to the European climate, for example), and global warming could affect certain authorizations in the long term. In addition, French regulations may be partially or completely taken up by Europe in the long term. This path is indeed currently followed by the EU with recent regulatory incentives (EU 2021a) paving the way for the regulation of macroorganisms.

1.2.3. *Limits with crop protection in Organic Agriculture*

First of all, the four pillars of biocontrol are acceptable in Organic Farming ruled by Regulation EU 2018/848 (EU 2018b), the question of GMOs or GMO-produced substances being regulated in a similar way to the previous regulation (EU 2008b). The current global restriction avoiding the use of herbicides (a de facto total ban) implies that weeds are managed mainly by physical methods (mechanically, thermically or electrically). The global question of crop protection is managed primarily by protection from natural predators, the choice of species and varieties, crop rotation, cultivation techniques, mechanical methods and physical and thermal processes. Questions on new herbicide techniques (electricity) are currently being assessed. The question of "transformed" natural substances is also at the heart of debates. Three out of four pillars are automatically authorized in Organic Production: macroorganisms, microorganisms and chemical mediators. The authorizations of natural substances (of plant, animal, mineral and microbial origin) are managed on an ad hoc and individual basis for each active substance and are subject to voluntary inclusion files in Annex I (protection of crops and foodstuffs and stored products by extension), deposited at the national level and managed by the DGAgri through EGTOP PPP examination mandates (EGTOP 2017). Most of these natural biocontrol substances are thus accepted (Marchand 2017), except herbicides and substances that regulate positive (auxins) or negative (herbicides) growth. Annex I is under deep discussion, with the creation of sub-parts that did not previously exist, such as natural substances obtained from microorganisms (currently spinosad, cerevisane, ABE IT 56), specific parts for low-risk substances (EU 2017b; Robin and Marchand 2021b, 2021c) and basic substances (Marchand 2015, 2017; Robin and Marchand 2019a; Marchand et al. 2021). In conclusion, almost all biocontrol substances and products of the four pillars are accepted in Organic Agriculture (EGTOP 2021).

1.2.4. *Limits of biocontrol: contentious substances!*

Natural substances, including some chemical mediators, are not free from toxicity, and many examples show this, such as rotenone. Aside the natural substances historically used for lethal purposes (Bacalexi and Katouzian-Safadi 2018), some natural substances have been withdrawn from approval (e.g. crude tall oil) (EU 2017a), or never got approved in PPP Regulation (e.g. rotenone, nicotine) (EC 2008a). Direct comparisons show that the differences between “synthetic” and “natural” substances are not as sharp and the conclusions are sometimes not as dichotomous as society may think (Smith and Perfetti 2020). A similar acceptance problem may be encountered with copper compounds that are candidates for the substitution (as for emamectin, for example (EU 2020; Robin and Marchand 2021b)).

1.3. A mixed evolution

Biocontrol substances are increasing in number and percentage, relative to the total number of active substances approved in PPP regulation, mainly through the reduction of synthetic chemical substances (for the percentage) and subdivision of the lists of microorganisms by strain as an individual active substance (for the amounts). Of course, newly approved BCAs also account for the increase in both indicators (amount and percentage). In addition, the delimitation between biocontrol and synthetic substances is unclear, because certain synthetic chemicals can be used in biocontrol, in particular, in the "kill" part of "attract and kill" traps, which are also approved in organic farming plant protection (EU 2021b), although some member states asked to remove them from Annex I (Marchand 2019).

Thus, the number of biocontrol substances rose from 147 (38%) initially in 2011 to 216, representing 48% of all active substances, in 2022. The increase is very significant in number (+ 47%) but still low in overall percentage. The date by which biocontrol substances will represent 50% of all active substances, initially predicted to be 2025 (in 2018), is currently scheduled for 2022 – which is today. It is important to reason in terms of the active substance because a single substance can give rise to many different products, usages and even functions. As such, the visible increase in the number of biocontrol products on the French List only comes from the multiplication of products, often for redundant uses, based on just a few substances. Sometimes, even new active substances only have redundant uses, like ferric pyrophosphate and the already-approved ferric phosphate (Robin and Marchand 2021b).

In spite of this encouraging growth, the arrival of new active biocontrol substances and solutions approved under the PPP regulation is extremely slow, even if the annual percentage of new biocontrol substances in relation to the total annual new active substances is increasing (and was even 100% in both 2020 and 2021). The approval of biocontrol active substances submitted by petitioners is far from guaranteed, including for candidates with market authorizations in other countries (USA, Canada) ending up unapproved in Europe after the usual obstacle course of active substance approval (Vekemans and Marchand 2020).

Despite the regulatory changes since the implementation of Regulation 1107/2009 in 2011 (EC 2009a) compared to previous Directive 91/414/EC, progress is still slow. Even the renewal of these substances is not assured despite the interest they represent; many previously approved biocontrol substances have been lost, not supported for renewal or abandoned by petitioners for varying causes: too few uses and/or not enough specificity (estimated or real) leading to low sales volumes. In four years, 6% of the 80 substances lost were biocontrol substances. This phenomenon particularly affects substances that are accessible by other means than the phytosanitary protection market (pepper powder, ammonium phosphate, ammonium acetate). Reapproval of these active substances into basic substances is still possible. Basic substances (Marchand 2015) currently account for 11% of biocontrol substances (24 a.s. out of 216); they do not require market authorization and therefore are not protected. Nevertheless, a significant number of new microorganisms, 15 substances as of January 2022, are in the process of being approved, many with low-risk status.

1.3.1. *Struggling bases*

With regulatory considerations of the "new" PPP regulation arriving in 2011, low-risk substance rapidly appeared to be poorly defined since modifications occurred in 2018 (EU 2017b, 2018a) and new modifications for microorganisms are ongoing, hopefully for the better.

1.3.2. *Achieved progress*

A lightening of constraints has occurred in response to the difficulties encountered during the approval and renewal of microorganisms. Indeed, microorganisms as biocontrol active substances could legitimately claim the status of low-risk active substances. Previously, it took years of existence of the PPP regulation to modify the criteria for low-risk substances so that microorganisms could more easily access this status, whereas the previous criteria (DT50 less than 60 days) of point 5 of Appendix II strongly prevented microorganisms from achieving this status (EU 2017b). Appendix II of Regulation 1107/2009 was therefore seriously modified.

The Commission (DGSanté) thus organized a working group in 2017 to refine the criteria previously defined for low-risk substances. This work was subsequently published, updating Annex II, point 5, with Regulation (EC) No. 1107/2009, distinguishing between microorganisms and natural substances. This work also made it possible to establish a list of potential low-risk substances from already approved active substances (EU 2018a). As no direct transfer or grant from the defined list was recognized at that time, the transfer operation to low-risk status was deemed finalized during the individual renewal of the active substance. It is therefore slow and haphazard since one of the potential substances, garlic extract, has not been renewed in a low-risk substance (Robin and Marchand 2021c).

This statute must ensure marketing authorizations by law within 120 days of approval, but in fact, the deadlines are much longer (up to 754 days on average in Germany). In addition, the benefits granted to these low-risk substances depend on Article 22 (EU 2017c). Ongoing progress in the modification of microorganism evaluation, concretely “amending Annex II to Regulation (EC) No. 1107/2009 as regards specific criteria for the approval of active substances that are microorganisms” to consider all low-risk criteria applying to viruses which are non-virulent variants of plant pathogens, rather than to baculovirus only.

1.3.3. *Regulatory relief for semiochemicals*

1.3.3.1. *Features*

Semiochemicals are one of the four types of biocontrol and are considered active substances, thus falling under the 1107/2009 regulation (EC 2009a) when it comes to communal approbation.

These substances are emitted at very low levels in the environment by many different organisms. They fulfill a wide variety of roles and communications within and between species. It results that humans and other non-target species, as well as the environment, are constantly exposed to very low levels of various semiochemicals emitted by the different organisms present in the environment (insects, fragrant plants, etc.). As such, their mode of action is non-lethal and very specific to the receptor system of the receiving organism. These receptor systems are tuned to natural emission rates, which means that an efficient rate for plant protection use must be of the same magnitude. Indeed, a rate that is too high can modify or negate the expected effect. Because of these properties, it is expected that semiochemicals used as plant protection product pose a low risk to human health and the environment.

In the context of plant protection, different semiochemical communications have been exploited. The most common use is mating disruption, where a female sex pheromone is heavily released into the environment, which disrupts male ability to find a female, thus reducing mating and the next generation. Attractant pheromones can be used in “attract and kill” systems, in which an attractant is used to lure insects into a trap where they are killed by different means (insecticide, drowning, glue, etc.), or in "mass trapping" systems that use male emitted pheromones to attract and trap females, thus reducing egg-laying and subsequent larvae damage. Finally, repellents can be used to deter insects or inhibit their behaviors (feeding, ovipositing, etc.), although this last category of plant protection product has not yet been properly developed.

Guidelines. Straight Chain Lepidopteran Pheromones (SCLPs) representing a large part of the semiochemicals, a general "Guidance Document on the assessment of new substances falling into the group of Straight Chain Lepidopteran Pheromones (SCLPs) included in Annex I of Council Directive 91/414/EEC", was published in 2008 (DGSanté 2008). To further facilitate the application and approbation of semichemical substances, and to integrate the knowledge gained since 2003, a new "Guidance Document on semiochemicals active substance and plant protection products" was published in May 2016 (SANTE/12815/2014 rev. 5.2) (DGSanté 2016a) and later validated by OECD (OECD 2018) accompanied by efficacy evaluation guidance (EOPP 2019).

1.3.3.2. *Natural exposure level and reduced risk evaluation*

The new guidance document introduces the notion of natural exposure level, defined as "the level of exposure that might occur in the environment by high population of emitting organisms thus expected to be experienced by humans and other non-target organisms".

This notion is a key point in semiochemicals risk assessment. Indeed, if the applicant shows the use of their product results in environmental exposure by the same route and of lower or equivalent rate to the natural exposure level, then the risk assessment is concluded. In this ideal situation, only the identity, physical and chemical properties and information about the formulated product are needed. The sections about mammalian toxicology, residues and MRL, effects on non-target species and environmental fate and behavior do not have to be completed. The Guidance provides two different formulas to calculate the natural exposure level and exposure rate resulting from use as PPP.

However, if the exposure resulting from use as PPP is significantly higher than the natural exposure level then the final concentration derived from using the product should be calculated using a proposed model. The risk assessment continues, and all sections must be completed. Similarly, if oral and/or contact exposure is possible (e.g. spray, granules, treated seeds), then the full risk assessment to these routes is always needed.

1.3.3.3. *Natural exposure level and reduced risk evaluation*

Taking into account semiochemical specificity, the guidance document provides some advice on how to integrate the specific features of semiochemicals into the required dossier data.

As much detail as possible must be given about the biology and communication mechanisms between the organisms involved, about the specificity of that interaction and about the mode of action of the product in terms of how it modifies the behavior of the target organism. All this information on the biology of the interaction must be provided to justify the risk assessment strategy.

The use of semiochemicals as PPP relies on the release of low rates of volatile product for a long time, over different patterns, to maintain levels of exposure both tuned to the target species receptor system, and of the same magnitude as natural exposure level. The exposure rate is heavily determined by the characteristics of the dispenser or type of application, and as such, they must be thoroughly described. The guidance document provides details and examples in its appendix on the different types of existing application techniques. Moreover, information on the density, position, periods and frequencies of use of the dispensers must be provided, as well as factors affecting how the product should be used (weather, landscape, etc.).

All dossiers require demonstrating that the PPP is sufficiently effective and does not have unacceptable effects on the plant or plant products. The document notes that deviation from standard requirements and guidance may be required for some parts of the biological assessment dossier to account for the specific properties of semiochemicals. It is also noted that efficacy field trials for semiochemicals may be difficult to replicate and on a large scale. It is accepted that semiochemicals used as PPP may provide more or less control and have a more variable performance compared to a conventional product. Any reduced efficiency in terms of pest elimination will not in itself be a cause for refusal and other measures of efficacy will be considered, such as offering an alternative mode of action relevant to resistance management, reduced residues and compatibility with specific growing systems. Overall, the applicant must demonstrate a statistically significant improvement of an appropriate measure of either pest control, crop damage or crop yield that would be beneficial from an agronomic perspective.

1.3.3.4. *Specific case of attractants within traps*

1.3.3.4.1. EU implementation

In the Guidance document, it is specified that "semiochemicals are not considered active substances, when they are used only to attract arthropods" with the purpose of monitoring or within traps where they are subsequently killed. This means that when semiochemicals are attractant of arthropods used in traps, they do not fall under the 1107/2009 PPP regulation and are thus exempt from registration.

1.3.3.4.2. French implementation

On September 22, 2020, the French authority published a technical instruction (DGAL/SDQSPV/2020-581) reprising this notion. It clarifies those traps used for monitoring or "mass trapping" that rely on semiochemicals as attractant do not need market authorization if the trap is exempt of insecticides. The same exemption from market authorization applies if the attractant within the trap is a food product. On the other hand, if insects attracted to the trap are killed by an insecticide active substance, both the communal approval of the insecticide and the market authorization of the trap are required. This last technique, which avoids contact with chemical insecticides, is relatively safe and is also authorized in Organic Production (Marchand 2019).

1.3.4. *Regulatory relief for microorganisms*

1.3.4.1. *Features*

Microorganisms are one of the four types of biocontrol and are considered active substances, thus falling under the 1107/2009 regulation (EC 2009a) when it comes to communal approbation.

General considerations are described by "Guidance Document for applicants on preparing dossiers for the approval of a microbial active substance" published in March 2016 (SANTE/12545/2014–rev. 3) (DGSanté 2016b), but specific points may be found in "Guidance Document on the assessment of new isolates of baculovirus species already included in Annex I of Council Directive 91/414/EEC" published in 2008 (SANCO/0253/2008 rev. 2) (DGSanté 2008) and "Guideline developed within the Standing Committee on the Food Chain and Animal Health on the taxonomic level of microorganisms to be included in Annex I to Directive 91/414/EEC" published in 2005 (SANCO/10754/2005 rev.5) (DGSanté 2005).

1.3.4.2. *Evolution*

Considering the specificity of microorganisms, recent documents were generated by the Commission in 2020 (DGSanté 2020a, 2020b) in order to consider and manage metabolite production (SANTE/2020/12258) and antimicrobial resistance (SANTE/2020/12260) especially during low-risk status considerations.

1.3.5. *Regulatory relief for natural substances*

Natural substances are one of the four types of biocontrol and are considered active substances, thus falling under the 1107/2009 regulation (EC 2009b) when it comes to communal approbation. Botanicals, natural substances from plant origin, are the only one of the natural substances considered by approval guidance. The possible explanation is that only few natural substances of animal/microbial or mineral origin are candidates.

General considerations for plant extracts are described by "Guidance Document on botanical active substances used in plant protection products", published in March 2014 (SANTE/11470/2012 rev. 8) (DGSanté 2014).

1.4. Necessary evolutions

1.4.1. *At the EU level*

An effective first step towards a better consideration of biocontrol would be at least to include a clear definition of "biocontrol" in plant protection regulation (EC 2009), or even to create a specific category within the regulation to ease understanding. Although the approbation process and dossier size are reduced for biocontrol substances and relatively easier to compile compared to synthetic substances, this progress in promoting approbation of biocontrol substances only favors some of the four pillars. Thus, among recent applications, most are variations of microorganisms with similar behaviors: eight *Bacillus* sp. and six *Trichoderma* sp. Furthermore, the SUD (EC 2009b) should be an effective lever for the amplification of biocontrol, which it is not (Robin and Marchand 2019c, 2022; Vekemans and Marchand 2020, 2022).

1.4.2. *At the national level (France)*

All low-risk substances and almost all basic substances are biocontrol substances. The inscription of substances on the list of "potentially" low-risk substances could also be a positive sign. In contrast, reducing requirements does not seem to be on the agenda (Smith and Perfetti 2020) apart from classification by families, for example, *Bacillus* sp. Meeting deadlines seems an essential prerequisite for the intensification of substances into biocontrol products, and the overall shortening of approvals. Clearer and more accessible information, as well as training in the use of these products also seems necessary to promote the growth of this market. A national list of biocontrol products (with market authorization), now implemented with Organic Production status for each product, is accessible and also promotes BCAs.

Certiphyto permits. The Certiphyto certification for each individual councillor and user (farmer) delivered under the SUD (EC 2009b) is an acknowledgment/agreement for each holder of this certificate/permit to use pesticides. Even if the currently applied regulations show that the current orientation is rather towards the absence of Certiphyto obligation for these products (biocontrol), this is in complete contradiction with the technical knowledge actually needed to use them properly. The introduction of a specific "Biocontrôle" Certiphyto permit could provide for this a specific required knowledge for these BCAs.

1.5. Conclusion

BioControl Agents are indeed globally less invasive means of plant protection, including those from the biopesticide type. All low-risk and almost all basic substances (except hydrogen peroxide, although it is involved in natural defense mechanisms) are indeed BCAs, but this constantly growing part of plant protection active substances is still facing criticism regarding the few remaining litigious natural substances, including few candidates for substitution. However, although regulatory approval pathway was updated and technically modernized, a lot of promising BCAs are still rejected by the process, compromising the speed of implementation of biorational/IPM implementation. The latest regulatory changes in progress should only impact the "low-risk" status of new BCAs and only slightly increase the number of BCA substances, which have only averaged five to six BCA active substances per year since 2011.

1.6. References

Bacalexi, D. and Katouzian-Safadi, M. (2018). Drug expertise and drug experts in antiquity, oriental middle ages and occidental renaissance: Knowledge, know-how and social impact. *Scientiae Disciplines of Knowing in the Early Modern World, Scientiae International Research Group*, Minneapolis, May, 1–9.

Charon, M., Robin, D., Marchand, P.A. (2019). The major interest for crop protection of agrochemical substances without maximum residue limit (MRL). *Biotechnol. Agron. Soc. Environ.*, 23(1), 22–29.

DGSanté (2005). SANCO/10754/2005 rev. 5, Guideline developed within the standing committee on the food chain and animal health on the taxonomic level of microorganisms to be included in Annex I to Directive 91/414/EEC, 1–4.

DGSanté (2008). SANCO/0253/2008 rev. 2, Guidance document on the assessment of new isolates of baculovirus species already included in Annex I of Council Directive 91/414/EEC, 1–21.

DGSanté (2010). SANCO/5272/2009 rev. 3, Guidance document on the assessment of new substances falling into the group of Straight Chain Lepidopteran Pheromones (SCLPs) included in Annex I of Council Directive 91/414/EEC, 1–29.

DGSanté (2014). SANTE/11470/2012 rev. 8, Guidance document on botanical active substances used in plant protection products, 1–28.

DGSanté (2016a). SANTE/12815/2014 rev. 5.2, Guidance document on semiochemical active substances as plant protection products, 1–37.

DGSanté (2016b). SANTE/12545/2014 rev. 3, Guidance document for applicants on preparing dossiers for the approval of renewal of approval of a microorganism including viruses according to Regulation (EU) No. 283/2013 and Regulation (EU) No. 284/2013, 1–9.

DGSanté (2020a). SANTE/2020/12258, Guidance on the risk assessment of metabolites produced by microbial active substances as plant protection products, 1–42.

DGSanté (2020b). SANTE/2020/12260, Guidance on the approval and low-risk criteria linked to antimicrobial resistance for microbial active substances applicable to microorganisms used for plant protection in accordance with Regulation (EC) No. 1107/2009, 1–14.

EC (2008a). 2008/317/EC: Commission decision of 10 April 2008 concerning the non-inclusion of rotenone, extract from equisetum and chinin-hydrochlorid in Annex I to Council Directive 91/414/EEC and the withdrawal of authorisations for plant protection products containing these substances. (C(2008) 1293), OJ L 108/30.

EC (2008b). Commission Regulation (EC) No. 889/2008 of 5 September 2008 laying down detailed rules for the implementation of Council Regulation (EC) No. 834/2007 on organic production and labelling of organic products with regard to organic production, labelling and control, OJ L 250/1–84.

EC (2009a). Regulation (EC) No. 1107/2009 of the European Parliament and of the Council of 21 October 2009 concerning the placing of plant protection products on the market and repealing Council Directives 79/117/EEC and 91/414/EEC, OJ L 309/1–50.

EC (2009b). Directive 2009/128/EC of the European Parliament and of the Council of 21 October 2009 establishing a framework for Community action to achieve the sustainable use of pesticides, OJ L 309/71.

EGTOP (2017). Commission decision of 30 August 2017 appointing the members of the group for technical advice on organic production and drawing up the pool list. C/2017/5767, OJ C 287/3.

EGTOP (2021). Final report on Fertilisers IV and Plant Protection Products VI. Directorate-General for Agriculture and Rural Development. EGTOP/2021, 1–25.

EOPP (2019). PP 1/264 (2) Principles of efficacy evaluation for mating disruption pheromones. *EPPO Bull.*, 49(3), 426–430.

EU (2017a). Commission Implementing Regulation (EU) 2017/1186 of 3 July 2017 withdrawing the approval of the active substance repellents by smell of animal or plant origin/tall oil crude, in accordance with Regulation (EC) No. 1107/2009 of the European Parliament and of the Council concerning the placing of plant protection products on the market, and amending Commission Implementing Regulation (EU) No. 540/2011. C/2017/4443, OJ L 171/131.

EU (2017b) Commission Regulation (EU) 2017/1432 of 7 August 2017 amending regulation (EC) No. 1107/2009 of the European Parliament and the council concerning the placing of plant protection products on the market as regards the criteria for the approval of low-risk active substances, C/2017/5386, OJ L 205/59.

EU (2017c). DG Health and Food Safety. Overview report on a series of audits carried out in EU Member States in 2016 and 2017 in order to evaluate the systems in place for the Authorisation of Plant Protection Products. DG(SANTE) 2017-6250–MR, 1–30, doi:10.2875/580084.

EU (2018a). Commission notice concerning a list of potentially low-risk active substances approved for use in plant protection. C/2018/4828, OJ C 265/8.

EU (2018b). Regulation (EU) 2018/848 of the European Parliament and of the Council of 30 May 2018 on organic production and labelling of organic products and repealing Council Regulation (EC) No. 834/2007, PE/62/2017/REV/1, OJ L 150/1–92.

EU (2020). Commission Implementing Regulation (EU) 2020/1295 of 16 September 2020 amending Implementing Regulation (EU) 2015/408 as regards the inclusion of the active substances carbetamide, emamectin, flurochloridone, gamma-cyhalothrin, halosulfuron-methyl, ipconazole and tembotrione in the list of candidates for substitution, C/2020/6239, OJ L 303/18–19.

EU (2021a). Council Decision (EU) 2021/1102 of 28 June 2021 requesting the Commission to submit a study on the Union's situation and options regarding the introduction, evaluation, production, marketing and use of invertebrate biological control agents within the territory of the Union and a proposal, if appropriate in view of the outcomes of the study, ST/9112/2021/INIT, OJ L 238/81–83.

EU (2021b). Commission Implementing Regulation (EU) 2021/1165 of 15 July 2021 authorising certain products and substances for use in organic production and establishing their lists. C/2021/5149, OJ L 253/13–48.

Journal Officiel "Lois et Décrets" (JORF) (2022). L. 253-6 du code rural et de la pêche maritime comme des agents et des produits utilisant des mécanismes naturels dans le cadre de la lutte intégrée contre les ennemis des cultures.

Marchand, P.A. (2015). Basic substances: An approval opportunity for low concern natural products under EU pesticide regulation. *Pest Manag. Sci.*, 71(9), 1197–1200.

Marchand, P.A. (2017). Basic and low risk substances under EU pesticide regulation: A new choice for biorationals portfolio of small and medium-sized enterprises. *J. Plant Prot. Res.*, 57(4), 433–440.

Marchand, P.A. (2018). Novel plant protection regulation: New perspectives for organic production? *Org. Farming*, 4(1), 3–6.

Marchand, P.A. (2019). Synthetic agrochemicals: A necessary clarification about their use exposure and impact in crop protection. *Environ. Sci. Pollut. Res.*, 26(18), 17996–18000.

Marchand, P.A., Davillerd, Y., Riccioni, L., Sanzani, S.M., Horn, N., Matyjaszczyk, E., Golding, J., Roberto, S.R., Mattiuz, B.-H., Xu, D. et al. (2021). Basics, an Euphresco international network on renewable natural substances for durable crop protection products. *Chron. Bioresour. Manag.*, 5(3), 77–80.

OECD (2018). Guidance document on semiochemical active substances and plant protection products. *Series on Pesticides*, ENV/JM/MONO(2017)33, 93, 1–48.

Robin, D.C. and Marchand, P.A. (2019a). Evolution of regulation (EU) No. 540/2011 since its entry into force. *J. Regul. Sci.*, 7, 1–7.

Robin, D.C. and Marchand, P.A. (2019b). Biocontrol active substances: Evolution since the entry in vigour of Reg. 1107/2009. *Pest Manag. Sci.*, 75(4), 950–959.

Robin, D.C. and Marchand, P.A. (2019c). Evolution of Directive (EC) No. 128/2009 of the European Parliament and of the Council establishing a framework for Community action to achieve the sustainable use of pesticides. *J. Regul. Sci.*, 7, 1–7.

Robin, D.C. and Marchand, P.A. (2020). Macroorganismes réglementés en France. *Innov. Agron.*, 79, 425–39.

Robin, D.C. and Marchand, P.A. (2021a). The slow decrease of the active substances candidates for substitution in the framework of the European Pesticide Regulation (EC) No. 1107/2009. *Eur. J. Risk Regul.*, 13, 1–22.

Robin, D.C. and Marchand, P.A. (2021b). Low-risk substances, new effective category of biocontrol agents as lever for durable crop protection products. *Chron. Bioresour. Manag.*, 5(1), 9–11.

Robin, D.C. and Marchand, P.A. (2021c). Evolution of the low-risk substances in the framework of the European Pesticide Regulation (EC) No. 1107/2009. *Eur. J. Risk Regul.*, 13, 1–20.

Smith, C.J. and Perfetti, T.A. (2020). A comparison of the persistence, toxicity, and exposure to high-volume natural plant-derived and synthetic pesticides. *Toxicol. Res. Appl.*, 4, 1–45.

Stenberg, J.A., Sundh, I., Becher, P.G., Christer, B., Mukesh, D., Egan, P.A., Friberg, H., Gil, J.F., Jensen, D.F., Mattias, J. et al. (2021). When is it biological control? A framework of definitions, mechanisms, and classifications. *J. Pest. Sci.*, 94, 665–676.

Vekemans, M.-C. and Marchand, P.A. (2020). The fate of the biocontrol agents under the European Phytopharmaceutical Regulation: A hindering for approval botanicals as new active substances? *Environ. Sci. Pollut. Res.*, 27(32), 39879–39887.

Vekemans, M.-C. and Marchand, P.A. (2022). The European pesticides harmonised risk indicator 1: A clarification about its displayed rendering. *J. Regul. Sci.*, 10(1).

2

Biological Controls in Horticulture

Nicolas GUIBERT[1], Maïder SAINT JEAN[2], Claire PRIGENT-COMBARET[3], Jean-Marc DEOGRATIAS[1], Cécile CABASSON[4] and Pierre PÉTRIACQ[4]

[1] *ASTREDHOR SUD-OUEST, France*

[2] *UMR 6060, Bordeaux University, France*

[3] *EM, UMR 5557, Lyon 1 University, France*

[4] *BFP, UMR 1332, Bordeaux University, France*

2.1. Introduction

2.1.1. *Horticulture in Europe*

Horticulture is the branch of plant agriculture that concerns fruits, vegetables and ornamental plants. Horticulture is divided into two categories: the cultivation of food plants and ornamental plants. This industry is generally very specialized in relation to its cultivation. As the product is perishable, transport to market must be organized to avoid losses (Mellington Synge et al. 2019).

The United Nations (UN) has declared 2021 as the "International Year of Fruit and Vegetables". The aim is to promote the nutritional qualities and health benefits of increased consumption of fruit and vegetables. The production of fresh fruit and vegetables in the EU represents 9% of the total produced worldwide in 2019, which amounted to 1,837 million tons.

Biocontrol of Plant Disease,
coordinated by Claire PRIGENT-COMBARET and Bernard DUMAS. © ISTE Ltd 2022.

France is the third-largest producer of fruit and vegetables in the European Union, behind Italy and Spain. Fruit and vegetables (including potatoes) represent 11% of French agriculture in value (FEPEX 2020).

Concerning ornamental plants, since 2017 the European Union represents one of the highest flower production densities per hectare in the world (10% of the world area and 44% of the world production of potted plants and flowers (see FranceAgriMer 2019)). The main countries involved in this production are the Netherlands with 32%, Germany, Spain and France equally with 12% and Italy with 11%. Therefore, the European Union had in 2017 a trade surplus in live plants and floriculture products of 327 million euros.

Today, the French sector is mobilizing to meet the expectations of consumers/citizens who are interested in different plant functions and uses (well-being, aesthetics, agri-food, health), and users of plants that reconcile the resilience of territories and the environment. For this, the use of biocontrol and biostimulants is the main lever to be mobilized (Piasentin 2009).

2.1.2. *Biocontrol solutions in horticulture*

Biocontrol (BC) methods are at the heart of the changes in practice that are taking place in European horticulture. BC has been known for a long time, and in 1967, Mr Koppert sold his first predatory mites in the Netherlands. Fifty years later, Koppert Biological Systems has become a significant supplier for vegetable crops and many ornamental plants in greenhouses (van Der Ent et al. 2018). Specialized crops have been pioneers in this field.

The ornamental horticulture sector took an early interest in biocontrol methods because of their specific characteristics (few approved products, confined spaces, permanent work activities in greenhouses, great diversity of cultivated species with all their pests and diseases, etc.). In France, as early as the 1990s, the ASTREDHOR technical institute launched important research work on Integrated Biological Protection. Producers quickly followed and now have good expertise in the field of biological protection (Astredhor 2018).

Since 2002, BC has been supported at thanks to the development of the "pesticides package" comprising Regulation (EC) No. 1107/2009 and Directive 2009/128/EC. The former establishes the rules concerning the authorization of the sale, use and control of plant protection products. To go further, directive 2009/128/EC seeks to develop alternative means of control in the field. It is then up to each state to propose an action plan to achieve these objectives.

However, Europe does not provide a definition of BC (see Chapter 1). In France, the Rural and Maritime Fishing Code defines BC as "agents and products using natural mechanisms in the context of integrated pest management". They include, in particular:

– macroorganisms;

– plant protection products comprising microorganisms, chemical mediators such as pheromones and kairomones and natural substances of plant, animal or mineral origin (Article L.253-6). Microorganisms are also referred to as microbial biological control agents and have been recently reviewed (Köhl et al. 2019).

In order to meet European demands, France launched the Ecophyto plan in 2008 following the Grenelle Environment Forum. Thus, specialized crops have 23 Dephy Expe programs that started in 2018 or 2019 out of 41 projects. This is more than half, which clearly illustrates the dynamics at work within the horticultural sector (Emonet and Rougier 2019).

The evolution of the BC market is monitored by the French association of BC product companies IBMA France. According to this association (IBMA France 2020), BC represented 11% of the plant protection market in 2019. Sales of these products amounted to 217 million euros, an increase of 8.5% compared to 2018. IBMA will target 30% of the market in 2030. France has a strong desire to develop BC and could become the driving force in Europe. The challenge is to support the Government in the European definition of BC, based on the French framework (*Journal officiel de la République Française* 2018). Moreover, in a resolution of February 15, 2017, the European Parliament stressed the need to revise European legislation to promote the development of low-risk pesticides of biological origin. This recognition would be consistent with the Green Deal guidelines for Europe (Hulot and Hiller 2021).

2.1.3. *Biostimulants, complementary to biocontrol*

Strict EU environmental safety regulations and support for organic farming have paved the way for a thriving biostimulants market in countries such as Germany, Italy and France. In its "Economic Overview of the European Biostimulants Market", the European Biostimulants Industry Council website reports that plant biostimulants can increase yields by 5–10% while complementing BC solutions. On June 25, 2019, the EU Regulation 2019/1009 was published in the Official Journal of the European Union, a harmonized regulation for all fertilizers and growing media that will come into force on July 16, 2022, finally giving a framework to biostimulants. There is no longer any risk of confusion with plant protection products, as the boundary is clearly defined around abiotic and biotic stresses (Figure 2.1).

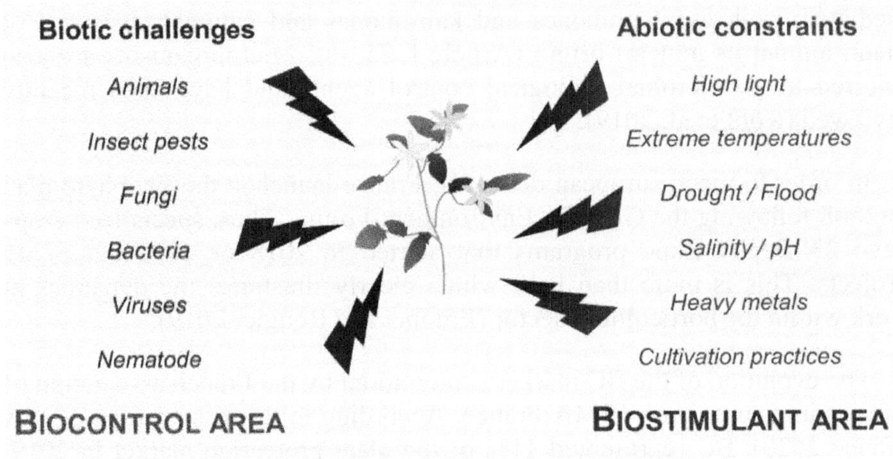

Figure 2.1. *Biocontrol and biostimulants can help plants cope with biotic and abiotic stresses*

A plant biostimulant is a product that stimulates plant nutritional processes independently of the nutrients it contains, for the sole purpose of improving one or more of the following characteristics of plants or their rhizosphere:

– nutrient use efficiency;

– tolerance to abiotic stress;

– quality characteristics;

– availability of nutrients confined to the soil or rhizosphere.

There are six groups of products according to their active ingredient: humic acid, fulvic acid, amino acids, protein hydrolysates, seaweed extracts and other active ingredients (including microorganisms). The "biostimulant" market is difficult to define because today biostimulants are not distinguished from fertilizing materials. Nevertheless, there is a constant increase of 9–10% per year in France and Europe and specialized crops are the main driver (Largant 2019).

2.1.4. *A biocontrol example: TRIANUM® P and G approval in France*

TRIANUM®-G and TRIANUM®-P are the first commercial products to be registered with *Trichoderma harzianum* thanks to the collaboration between the Dutch company Koppert Biological Systems (KBS) and experimental stations such as Astredhor Sud-Ouest GIE Fleurs et Plantes (Lacordaire et al. 2009).

First, trials were conducted in 2003 and 2004 at the Astredhor Seine-Manche on pansies. They highlighted the effect of TRIANUM® as a protective shield for the root system, statistically more efficient than the chemical reference against soilborne diseases.

The DRAF-SRPV and FREDON des Pays de la Loire conducted trials on hybrid Rhododendrons between 2005 and 2007. They demonstrated that the addition of *Trichodema harzianum* strain T22 to the substrate resulted in a better rooting rate of the cuttings, a more developed root system of the young plants and higher viability of the plants in pots (Jullien 2009).

Astredhor Sud-Ouest GIE Fleurs et Plantes also worked in collaboration with KBS from 2005 to 2015 to test the effectiveness of this strain on different plant/pathogen pairs. The microorganism was able to confirm its biostimulant action by promoting the recovery of cuttings within the framework of vegetative propagation, by promoting root development (increased root biomass and favored transplantation) and by also having an impact on the above-ground biomass of plants. However, the protective effect of the fungus against soilborne diseases was much more difficult to reveal. This highlights the importance of knowing the *Trichoderma*/pathogenic fungus/plant interaction well (Harman 2006).

Finally, the T22 strain remains the best known and TRIANUM® is still the only product registered in France based on *Trichoderma harzianum*.

2.2. Biological controls in horticulture

As previously mentioned, biological approaches for the control of microbial pathogens and pests have been integrated into horticultural practices over the past 40 years (Wilson 1997; Stenberg et al. 2021). First, research efforts were invested in the development of BC agents to modulate the foliar diseases caused by bacterial pathogens, control the soilborne fungal pathogens and regulate pest population in crops. During this period, BC agents were empirically selected to antagonize a single pathogen/pest.

From the beginning of the 21st century, the frequent occurrence of bacterial and fungal resistance to pesticides among horticultural productions and the societal awareness of the impact of pesticides on the environment and health have led to the development of several novel approaches (Pétriacq et al. 2018). They also ensure a more rapid integration of commercial BC products into horticultural practices. These novel approaches include (i) mixture of BC agents; (ii) integration of biological and chemical agents; (iii) integrated biological control strategies; (iv) rhizobacterial-mediated systemic induced resistance; (v) beneficial ecosystem regulation to sustain a genetic and functional diversity of the microbial community or natural predators and parasitoids. Since the 1990s, many BC classes have been successfully tested to control diseases and pests on horticultural crops or induce systemic resistance in the horticultural plants (Table 2.1 and references therein).

Evaluation of a wide range of products, putative host resistance inductors or biostimulants from various classes (e.g. mineral, organic, microorganisms) were made in certain species such as *Olea europaea* to control Verticillium wilt of olive caused by the hemibiotrophic soilborne fungus *Verticillium dahliae* Kleb under controlled conditions (López-Moral et al. 2021) or in the field (Mulero-Aparicio et al. 2020). Unfortunately, our understanding of the mechanisms of action in horticultural plants is still fragmentary. Indeed, most studies conducted in vitro showed that some BC agents produced a variety of antifungal secondary metabolites and lytic enzymes (Mérillon and Ramawat 2020). For instance, 2.4-diacetylphloroglucinol produced by *Pseudomonas fluorescens* strain F113 and lytic enzymes such as chitinase and protease from *Stenotrophomonas maltophilia* strain W81(P) were identified as biocontrols of the fungus *Pythium ultimum*, an agent of damping-off and root rot diseases for many ornamental species (Dunne et al. 1996). More recently, the role of elicitors has been ascertained in triggering disease resistance mechanisms in plants. In particular, salicylic acid (SA), a plant defense hormone, harbors fungistatic properties on the mycelial growth of *Eutypa lata*, the causal agent of *Eutypa* dieback, a severe xylem infection of woody plants (Amborabé et al. 2002). In conditions mimicking the plant environment (pH 5.5), 2 mM SA affected the sub-cellular organization of mycelia. The polypeptide fraction isolated from the culture medium of mycelium induced an early response on plant cell membrane with an increase of H+ conductance, activation of NADPH oxidase and phenylalanine ammonia lyase (Octave et al. 2006). Besides, H+-ATPase was confirmed to play a major role after the application of chitosan, a polymer of ß-1,4-glucosamine residues (deacetylated derivative of chitin) with antifungal properties (Amborabé et al. 2008). This membrane depolarization leads to inhibition of the co-transport of metabolites such as sucrose and valine and the turgor reaction of motor cells. The benefits that many species of endophytic fungi, which deliver bioactive metabolites with antimicrobial, insecticidal or plant biostimulant activities, can offer to the host plant have recently been reviewed (Sinno et al. 2020).

One salient example resides in the evolution of the research investigation on the BC *Trichoderma sp.* On the negative side, *Trichoderma* could be pathogenic to pepper and cucumber seedlings (Marín-Guirao et al. 2016). On the contrary, mycorrhizal inoculation with *Trichoderma harzianum* on bedding plant species propagated from cuttings significantly increased dry weight and total shoot length of *Verbena* and *Torenia* plants (Šrámek et al. 2000). *Trichoderma harzianum* also showed a significant positive effect on the growth and flowering of cyclamen and inoculation substantially decreased plant mortality caused by spontaneous infection by the fungal pathogen *Cryptocline cyclaminis* (Dubský et al. 2002). These different results indicate that the study of trade-offs between growth and protection from pathogens is becoming central to understanding plant organisms and their plasticity of tolerance/resistance.

Biocontrol target	Plant/pathogen/biocontrol systems	References
Soilborne fungal pathogens	*Cucumis sativus/Pythium ultimum/Bacillus subtilis*	Batinic et al. (1998)
Bacterial pathogens	Fresh herb/*Escherichia coli*/Bacteriophage cocktail	O'Regan et al. (2019)
	Arabidopsis thaliana/Ralstonia solanacearum/Gluconacetobacter diazotrophicus (Plant growth promoting bacteria endophytes, PGPBEs)	Rodríguez et al. (2019)
Foliar fungal pathogens	Cucumber seedlings/*Pythium sp./Trichoderma harzianum*	Bell et al. (2000)
	Strawberry, cucumber, grape/*Botrytis cinerea/Trichoderma spp., Penicillium spp., Gliocladium roseum*	Wilson (1997)
	Cyclamen or Poinsettia/*Cryptocline cyclaminis/Trichoderma harzianum*	Dubský et al. (2002)
Powdery mildew	Grape/*Uncinula necator/Ampelomyces quisqualis*	Bodin et al. (2020)
Horticultural pests	Zucchini/Tomato leaf curl New Delhi virus vectored by tobacco whitefly *Bemisia tabaci*/predatory mite *Amblyseius swirskii natural enemie (Ne)*	del Mar Tellez et al. (2017)
	Flowering perennials plants/*Bemisia tabaci, Frankliniella occidentalis/Tobacco whitefly, western flower thrips*	Rodríguez et al. (2018)
Bio-induction of systemic resistance	Tomato/endophytic fungi	Reviewed in Sinno et al. (2020)

Table 2.1. *Some examples of biocontrol in horticulture*

2.3. Physiological trade-offs for growth and immunity

Plants are often facing various abiotic and biotic stresses that impact their growth. The activation of defense mechanisms at the expense of growth is known as the growth–defense trade-off concept (Figueroa-Macías et al. 2021). When perceiving pathogens or environmental changes, plants induce a rapid activation of defense mechanisms. This enhanced expression of defense responses uses resources that are no longer available for growth. Contrariwise, high plant growth and development hijack the use of resources allocated to plant defenses (Guo et al. 2018). There are different lines of defense in plants. The primary innate immune response in plants is activated by the detection of microbial- or pathogen-associated molecular patterns (MAMPs or PAMPs, respectively) that are recognized by pattern recognition receptors and is known as PAMP/MAMP-triggered immunity (PTI/MTI). The second plant innate immune response is triggered by the recognition of specific pathogen effector proteins and is referred to as effector-triggered immunity (ETI). All these lines of defense are ecologically costly. Thus, plants have to balance the costs and benefits of activating defense responses. In both PTI/MTI and ETI, several defense responses are observed similarly, such as calcium influx, reactive oxygen species burst, pathogen-induced callose deposition at the infection site and induction of the expression defense genes. A network of molecular signaling pathways is involved in the reprogramming of gene expression in order to get the appropriate physiological plant responses. Several phytohormones are commonly involved in plant defense and plant growth responses: salicylic acid, jasmonate, ethylene and abscisic acid are involved in stress responses while auxins, gibberellins or cytokinins, among others, are signaling molecules involved in growth and development. However, crosstalk occurs between these different regulators, often acting antagonistically, to obtain, in stressful environmental conditions, the best development of the plant (Eichmann and Schäfer 2015; Checker et al. 2018; Ballaré and Austin 2019; van Butselaar and Van den Ackerveken 2020).

Plants interact with a multitude of microorganisms forming their microbiota, the whole forming the plant holobiont. Within plant microbiota, some microorganisms will benefit the plant (mutualist symbionts, cooperative microorganisms) while others will negatively affect the growth of the plant (phytopathogens, phytoparasites) (Drogue et al. 2012). How is a plant able to distinguish microbial symbionts from microbial pathogens? Both can induce MTI responses, however, plant-beneficial microorganisms also prime an induced systemic response (ISR), while pathogens induce ETI in response to secreted virulence factors. Priming in the ISR response makes it possible to awaken defense responses in the plant, which, upon recognition of a pathogen, will lead to a faster and more effective response to subsequent attacks of pathogens (Hilker et al. 2016). All in all, the growth-defense trade-off will be in favor of growth when the plant encounters plant-beneficial microorganisms and defense when the plant faces pathogens or abiotic stresses. For instance, it has been shown that the association of plants with beneficial fungi like an *Epichloe* endophyte or certain *Trichoderma* may lower this trade-off (Hermosa et al. 2013; Bastías et al. 2021). This would be a consequence of the double ability of these endophytes to stimulate plant growth and protect the plant against biotic and abiotic stresses. It is interesting to note that plant defense responses could be inherited by offspring from parents through the inheritance of DNA methylation changes (Holeski et al. 2012).

The growth-defense trade-off concept suggests that fast-growing species will invest more available resources in plant growth rather than in defense mechanisms, whereas slower-growing species will conversely invest more resources in defense rather than in growth. Similarly, young tissues will reduce immune responses to maximize growth, whereas mature organs will invest in energy to defend themselves (Wang and Wang 2014).

During the green revolution, breeding programs were carried out to boost crop growth and yields in optimal environmental conditions (in the presence of high fertilization levels), but without specifically selecting plant genotypes for their ability to respond to stresses and to positively interact with plant-beneficial bacteria from soil (Valente et al. 2020). Hence, the genetic diversity for defense mechanisms against abiotic or biotic stresses could be counter-selected. Biocontrol agents and biostimulants could allow for the reduction of plant growth-defense trade-off, thereby leading to high crop growth and stress alleviation. In the face of climate change, the control of growth-defense trade-off becomes of prime importance.

2.4. Eco-innovations and economic trade-offs

Farmers play a dual role both as users of BC products and developers of new production methods, which leads them to face various economic trade-offs related to eco-innovation.

2.4.1. *Determinants of eco-innovation and trade-offs at the farmer level*

Eco-innovation is defined as the "production, assimilation or exploitation of a product, production process, service or management or business method that is novel to the organization (developing or adopting it) and which results, throughout its life cycle, in a reduction of environmental risk, pollution and other negative impacts of resource use" (Kemp and Pearson 2007). Environmental regulation is a key driver of eco-innovation, through a regulatory push/pull effect (Rennings 2000). However, eco-innovation cannot be considered as a direct and systematic response to regulation. The traditional determinants of innovation – namely the *technology push* and *demand pull* effects – also play a pivotal role (Kesidou and Demirel 2012). Eco-innovations are thus multifactorial and multidimensional by nature. They express technical-economic compromises by which firms seek to combine environmental objectives with objectives of competitiveness and cost (Oltra and Saint Jean 2005). Trade-offs between economic and environmental/health performances are often inevitable in the first place. The higher the barriers generally attached to radical eco-innovations (technological, financial, market, regulatory barriers[1]), the higher the trade-offs.

Eco-innovations in agriculture in general, and in horticulture in particular, encompass many environmental dimensions (energy use, carbon emissions, water, soil, pesticide use, nutrients) (Berkers and Geels 2011). They involve a combination of product, process and organizational innovations at different levels of the value chain. BC products represent a particular sub-type of eco-innovation adopted by farmers.

Such BC products rely on innovative principles "*to stimulate natural processes to enhance plant nutrient uptake, nutrient use efficiency, tolerance to abiotic stress, biocontrol, and crop quality*" (Hamid et al. 2021). Meanwhile, they reduce or eliminate the necessity to rely on chemical fertilizers and pesticides with adverse environmental and health impacts. The capacity to combine effectiveness in nutrition use, mitigation of abiotic/biotic stress, crop quality characteristics and yields or productivity improvement offered to farmers is a challenging and complex equation that BC products have to face to help create a more sustainable agriculture. Knowledge of mechanisms underlying those technologies and their multi-effects on plants is a necessary condition for their further development and adoption, especially to demonstrate their comparative advantages and the possibilities to escape from the pesticide lock-in.

1. Radical eco-innovations are generally associated with specific barriers: (i) higher technical complexity and/or incompatibility with the existing technologies; (ii) higher financial risk due to temporal arbitrage in favour of long-term effects; (iii) higher information asymmetry due to credence good characteristics of green products making consumers unable to objectively assess their environmental quality and unwilling to pay a premium price; (iv) higher dependence on policy design features.

According to data collected from the DEPHY FERME survey (CAN DEPHY 2018), conducted at the end of 2017 on 89 horticulture farmers, discovering alternatives to glyphosate in horticulture, almost 70% of the producers in the sample still use glyphosate for local applications or treatment of lanes and borders. Flowering potted plants and nurseries are the primary glyphosate users compared with producers of cut flowers who have managed to do without. The main barriers to glyphosate reduction reflect the conflict between a well-established substance and a bundle of less proven alternatives. The former benefits from the strong advantages in terms of the great efficiency of the molecule: its ease of application (large scope), low cost and long-lasting action. By contrast, natural herbicides appear unsuitable for a wide variety of crops; their usage is idiosyncratic[2], complex, time-demanding and short-acting. Moreover, new mechanization solutions are difficult to implement given the firm-specific localized needs that require a customized approach rather than a turnkey one.

Other studies confirm and extend the type of barriers under issue here. For instance, Havardi-Burger et al. (2020) examine the obstacles to eco-innovation for flowering potted plants for the German market through a series of interviews with actors located at different stages of the value chain (breeding and propagation nursery, grower, wholesaler, certifier, retailer) (Hasler et al. 2016). Main difficulties come from increased costs, financial risk and complexity of implementation. As well as this, the lack of product transparency prevents the transfer of sustainability costs to the consumer by offering a sustainable product for a premium price. Meanwhile, retailers have become more powerful chain actors, thus allowing them strategically use business-to-business standards to create an entry barrier for suppliers.

Those studies confirm that the diffusion of BC products depends on the ability of farmers to combine environmental performances with gains in productive efficiency and/or higher-quality products. For their part, consumers who become more demanding in terms of guarantees regarding the product quality, especially their environmental and health effects or local

2. Hernández-Rosas et al. (2020) state that the use of biocontrol remains limited by factors such as (i) the specificity of some microorganisms which control only one pest and not all those that can occur in a crop; (ii) they are not capable of having a significant effect on the control of the pest, or they only provide partial control or their efficacy is variable due to different climatic conditions; (iii) high production costs.

and sustainable sourcing, are facing goods with "*credence* characteristics"[3] when purchasing flowers and ornamental plants. Voluntary and/or mandatory labeling is thus necessary to help consumers make an informed choice.

Therefore, two related issues arise for BC products in horticulture. First, from a scientific and technical knowledge perspective, it is not always clear whether BC products have a positive or negative impact on all sustainability dimensions (variability of BC mode of action; insufficient proof of concept as to the economic value added). Second, from a consumer perspective, it is not easy to judge plant quality or to believe in claiming better environmental performance based on the mere appearance and aesthetic appeal of plants. The price/quality ratio is not enough to signal plants with differentiated process attributes. Certification and labeling, although necessary for orientating consumers' subjective beliefs, can have critical implications for the marketplace (market entry barrier, quality level of standard, price competition).

2.4.2. *Influence of sectoral innovation patterns*

Research in innovation studies at the sectoral level accounts for differences in the production and use of innovations between sectors. Those differences result from specific learning modes and equally specific sources of technological knowledge. One of the aims of the well-known taxonomy by Pavitt (1984) is precisely to capture such relations mapping "industry types" and industry dynamics. According to that taxonomy[4], agriculture enters into the category of "supplier-dominated firms" that face price-sensitive consumers. Those firms are generally small, with weak in-house R&D and engineering capabilities. Besides, their appropriation means are mainly non-technical and based on professional skills, aesthetic design, trademarks and advertising. Such characteristics (sources of technology, type of consumer and means of appropriation) thus contribute to

3. Consumers cannot tell the quality of the good they have purchased even after consuming or experiencing it.

4. Four groups of sectors have been identified: (i) "Supplier-dominated" sectors (e.g. textile, agriculture); (ii) "Specialized suppliers" (e.g. producers of industrial machinery and equipment); (iii) "Scale-intensive" sectors (e.g. automotive); (iv) "Science-based" industries (e.g. microelectronics, informatics, drugs and bioengineering).

reducing the costs of technology trajectories. From that perspective, supplier-dominated firms make only a minor contribution to their process or product technology. Most innovations come from suppliers of equipment and materials. Such a sectoral pattern of innovation in agriculture may influence both the enduring lock-in situation in the pesticide paradigm and the way BC products are developed.

In their study of innovation in the pesticide industry from 1930 to 1980, Achilladelis et al. (1987) clearly show the progressive movement of industrial concentration around a few large firms with a strong R&D tradition that made a considerable contribution to pesticide innovations throughout the considered period. The leading five firms (Bayer, Geigy, Dow, DuPont and ICI) accounted for nearly half of all the radical innovations; the leading 10 firms for almost two-thirds (in addition, BASF, Hoescht, Shell, American Cyanamid, Sumitomo). However, they also show the importance of the links between the research in these leading firms and academic research. The success of these top firms was related to their capacity to assimilate and develop the results of fundamental research in chemistry. Over the period, environmental and health considerations have played a role in directing pesticide innovations towards areas that were imposed by regulation (e.g. ban of organochlorine insecticides) or driven by market demand (e.g. greater attention given to biodegradable non-accumulating insecticides). What also is worth emphasizing is both the co-development between regulation and successive generations of products and the ability of incumbents to diversify in-house technologies so that "the use, development and regulation of chemical plant protection products have had a close relationship with the companies creating them" (Payne-Gifford et al. 2021).

The current global BC market tends to be "a fragmented market with the presence of various small and regional players operating in it" (IndustryARC 2020). In terms of market shares, the top players in 2018 were Certis (5.78%), BASF (5.34%), Sumitomo Corporation (4.89%), Marrone Bio Innovations (4.57%) and Bayer (3.65%). However, the same study raises the fact that "the greater market share of these players can be attributed to highly diversified product portfolio and a large number of acquisitions and agreements taking place". Thus, the big incumbent agrochemical companies have continuously adjusted their pesticide discovery strategy over time and have expanded into biologically derived treatments, allowing them ubiquity and a potential for regulatory capture in the transition management process towards sustainable agriculture.

In order to throw some light on the dynamics of BC innovations over the recent decades, we will examine patent data. We have classified BC products into four groups (microorganism, chemical mediators, natural substances and insect traps) such as listed under articles L.253-5 and L.253-7 of *Code Rural et de la Pêche Maritime* (CRPM) in France (Ministère de l'Agriculture et de l'Alimentation 2021). Figure 2.2 displays the respective number of business product names and active substances by the leading groups of BC products. In addition, circular diagrams account for internal variety and uneven distribution of active substances for microorganisms and natural substances.

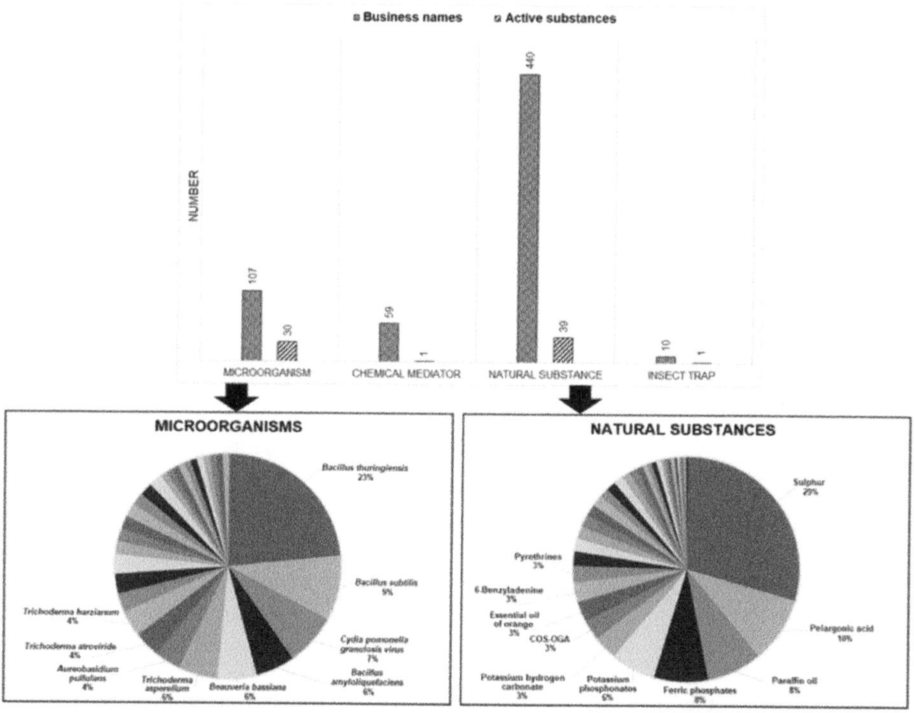

Figure 2.2. *Distribution of business names and active substances by leading groups of biocontrol products listed in Code Rural et de la Pêche Maritime*

For every group of BC, we clearly observe that the number of active substances is much more limited than the number of business product names. Microorganisms have a higher related variety (number of active substances in use/number of business names) compared with the three other groups. Within the two main groups – microorganisms and natural substances – we also observe internal disproportion with dominant active substances, respectively bacillus and sulfur, due to their broad spectrum of action and "generic purpose technology" characteristics.

Figure 2.3 presents cumulative curves of patent families for the period 1991–2018 for each group of BC products[5]. Figure 2.4 displays the main assignees for patent families in force.

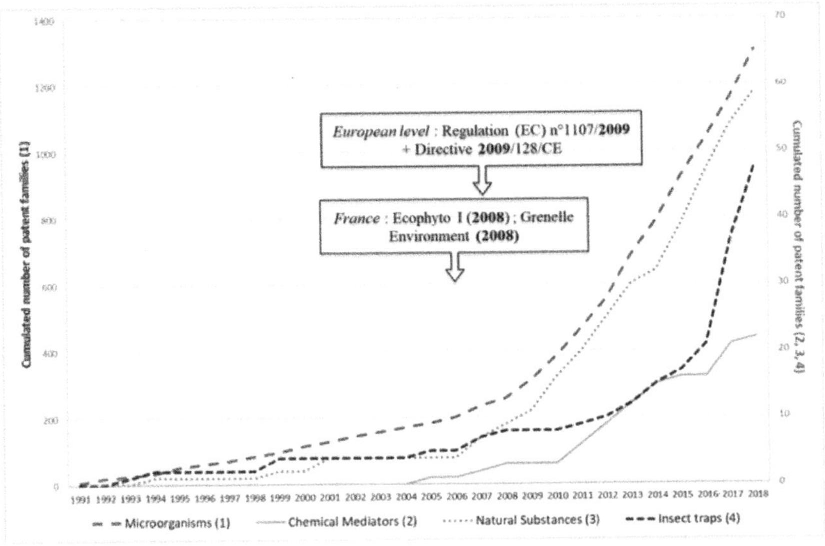

Figure 2.3. *Trends of patent families over the period 1991–2018*

5. The patent dataset is obtained by combining keyword searches on titles, abstracts and claims of patents and technological class searches, using the Orbit FamPat database (patents belonging to the same family are counted only once; world coverage of patent applications). The list of keywords corresponds to the list of active substances displayed in the CRPM. Four technological classes (IPC) belonging to the agriculture field are also selected (A01N, A01P, A01C and A01G) to restrict the search. We obtain a dataset of 1,712 patent families of which 1,168 are in force. A01N refers to "Preservation of bodies of humans or animals or plants or parts of thereof; biocides; pest repellents or attractants; plant growth regulators"; A01P to "Biocidal, pest repellent, pest attractant or plant growth regulatory activity of chemical compound or preparations; A01C to "Planting, sowing, fertilising" and A01G to "Horticulture; cultivation of vegetables, flowers, rice, fruit, vines, hops or seaweed; forestry; watering".

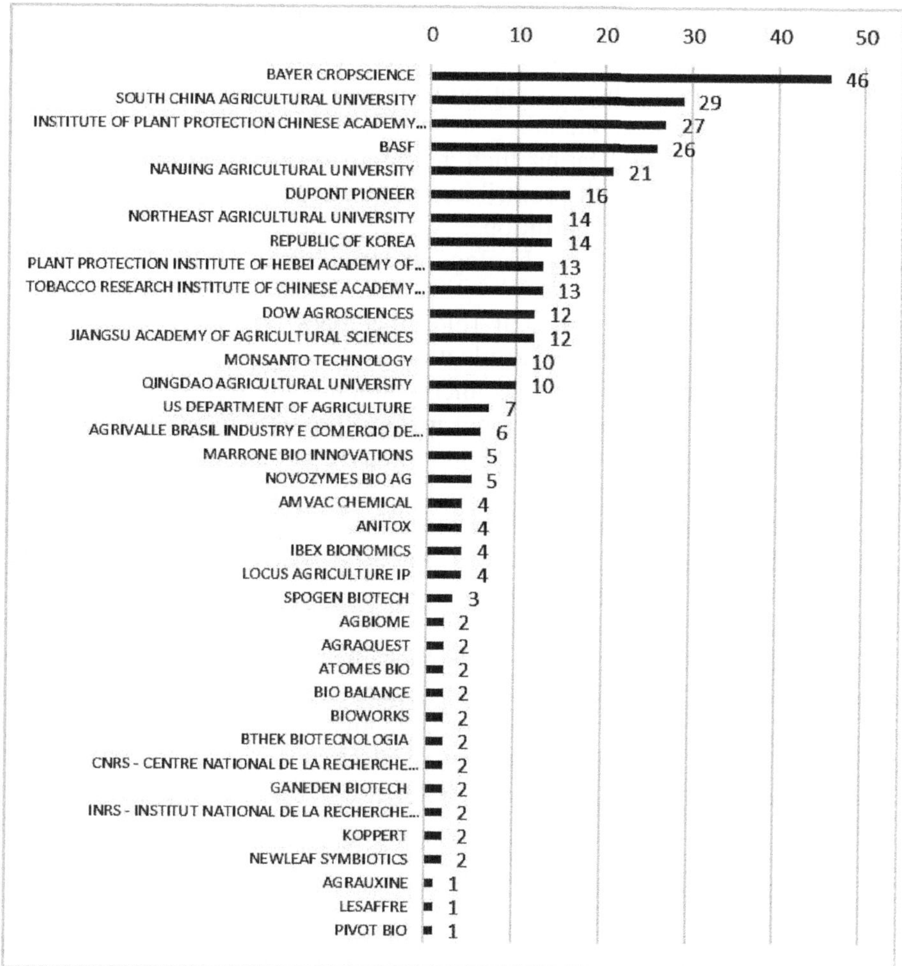

Figure 2.4. *Main patent assignees over the period 1991–2018*

Patenting curves show a clear take-off occurring between 2008 and 2010, suggesting a surge in BC product innovations at that time. This could reflect the coincidence of different factors (rising oil prices, growing concerns about pesticide exposure and soil depletion, stricter regulation) together with innovators and the search for substitutes. The analysis of assignees confirms the presence of incumbent agro-chemical firms with a solid and long R&D tradition in pesticides (Bayer, BASF, Dupont, Dow, etc.) as well as more recent and smaller or specialized firms (Agrauxine, De Sangosse, Koppert, Immunrise, Amoeba, De Sangosse, M2i Life Science, InVivo). The former have been diversifying their patent portfolios, well aware of the long-term potential of BC markets including a "watchdog" function. The latter are specialized suppliers dedicating their research to one specific type of BC product. We also observe the contribution of public and university research, especially from Chinese institutions. Suppliers and research thus tend to remain significant sources of technology in agriculture.

However, farmers play a dual role since they are both users of BC products and developers of new production methods to integrate them. Thus, restricting the role of farmers as mere adopters of product innovations ("supplier-dominated firms") does not make it possible to fully account for the fact that the development and adoption of BC products is likely to bring new learning modes and new sources of technological knowledge, giving rise (maybe) to a new sectoral pattern of innovation. The sectoral perspective highlights the dichotomy between R&D intensive and diversified firms at the heart of product innovations, on the one hand, and low-tech users (farmers) with idiosyncratic needs requiring organizational innovations (new methods and practices), on the other hand, (e.g. mechanical or thermal weeding, grass strips, mulching). Although the sector is characterized by supplier-dominated firms, the role of farmers as users and as innovators remains crucial to shift away from chemical pesticides to biological and/or practice-based ways of controlling pests (von Hippel 1988a, 1988b).

2.5. Challenges and perspectives

Remarkable efforts have been devoted to understanding the mode of actions of BC, which are essentially context-dependent. This is particularly true for microbial biological controls that harbor various interaction mechanisms, as thoroughly reviewed by Köhl et al. (2019). Briefly, biological controls "interact with plants by inducing resistance or priming plants without any direct interaction with the targeted pathogen" or "act via nutrient competition or other mechanisms modulation growth conditions for the pathogens" (Köhl et al. 2019). Antagonistic microbial BC could further directly interfere with the pathogenic microbe via hyperparasitism and antibiosis.

Nonetheless, further research is pressing to describe better how BC *largo sensu* interact with plants, including horticultural species. The rise of omics techniques in plant science is revolutionizing analytical strategies for studying BC and plant interactions. High-throughput sequencing methods have made it possible to identify the composition and function of the microbiome (Massart et al. 2015). More integrative approaches like metabolomics, which is not yet widely used, would complete our knowledge of BC-induced mechanisms (Mhlongo et al. 2018). Interestingly, artificial intelligence based on deep learning has also been deployed for horticultural research (Yang and Xu 2021). This promising approach will help promote intelligent horticulture by improving the phenotyping of various key traits for biological control.

Another main challenge is to make BCs more accessible to all, both technically and economically, as highlighted previously. For instance, the deployment of biological control solutions represents an exciting approach in urban horticulture for modern integrated pest management in urban areas (Feldmann and Vogler 2021). Cities could indeed offer great potential for horticultural food production as they gather essential socio-economic sectors for future city designs combining economic, ecological and societal demands (Edmondson et al. 2020). BC includes microbial agents and/or complex mixtures of molecules naturally present in the environment. It is thus assumed that BC methods offer environmentally friendly alternative strategies that have fewer undesirable side effects on ecosystems than conventional pesticides, such as for urban horticulture, where several challenges have been highlighted (Feldmann and Vogler 2021). However, some microbial BC might present possible pathogenicity under greenhouse conditions, thus leading to unsuitable contaminated horticultural products (Nicot 2011). In addition, BC methods also lead to the exposure to airborne microorganisms in the working environment of horticulturists and farmers (Tendal and Madsen 2011). Without a thorough study of the safety of these BC agents, it seems difficult to scale up the use of BC as alternative plant protection approaches.

Besides, there is increasing evidence of the compatibility of BC with pollinators, which have become more widespread in agriculture and horticulture in recent years. This raises the question of the potential risks of

BC against these insects. Two commercial compounds were tested in this context, namely, Binab-TF-WP and Binab-TF-WP-Konc, both of which contain combinations of *Trichoderma harzianum* ATCC20476 and *Trichoderma polysporum* ATCC20475 (Mommaerts et al. 2008). This study demonstrated that under laboratory conditions, the two BC treatments tested were safe for use in combination with *Bombus terrestris*.

Another aspect that deserves attention is the reproducibility of the results and outputs obtained from BC methods (personal communication, Astredhor 2021). Indeed, it would appear that the effectiveness of BC is not very reproducible due to the numerous biotic and abiotic factors involved in the plant–pathogen interaction. Annual variations can be observed in connection with modulations of the pathogen virulence, plant fitness, climatic or irrigation conditions, etc. Again, the general consensus calls for further research into these different challenges.

2.6. Concluding remarks

By 2030, the European Green Deal strategy aims to reduce the use of conventional pesticides by 50% and pesticides by 25% (Hulot and Hiller 2021). Biocontrol can be seen as a key element in achieving the European Green Deal targets. Mostly of natural origin, biological controls are products that meet the expectations of consumers looking for natural solutions. As such, BC products will be one of the main levers to protect crops, including horticultural species, and meet consumer expectations. While intensive efforts have identified some crucial, context-dependent modes of action for crops, further research is needed to increase our knowledge of how BCs interact with plants in general and particularly horticultural species. This research will, among other things, make it possible to gain reproducibility to define the best BC treatments, or a combination of treatments, to manage phytopathogenic diseases of horticultural plants. It is also essential to increase our knowledge of potential biocontrol resistant pests, and to do this, it is necessary not only to develop but also to implement tools and methods to assess and modulate this risk. Learning experiences will add to the general knowledge base of BC solutions and the benefits of use will increase the more farmers adopt them and are confident about their value for money.

In summary, there is general agreement that additional investigations are becoming mandatory to provide scientific evidence on the fate and behavior of biocontrol products in the environment in order to ensure the protection of the environment, to support the sustainability of these innovative products or to encourage and accelerate the introduction of new biocontrol products.

2.7. References

Achilladelis, B., Schwarzkopf, A., Cines, M. (1987). A study of innovation in the pesticide industry: Analysis of the innovation record of an industrial sector. *Res. Policy*, 16(2–4), 175–212.

Amborabé, B.-E., Fleurat-Lessard P., Chollet, J.-F., Roblin G. (2002). Antifungal effects of salicylic acid and other benzoic acid derivatives towards Eutypa lata: Structure–activity relationship. *Plant Physiol. Biochem.*, 40(12), 1051–1060.

Amborabé, B.-E., Bonmost, J., Fleurat-Lessard, P., Roblin, G. (2008). Early events induced by chitosan on plant cells. *J. Exp. Bot.*, 59(9), 2317–2324.

Astredhor (2018). *Guide technique sur les méthodes alternatives en protection des plantes*. Astredhor Editions, Paris.

Ballaré, C.L. and Austin, A.T. (2019). Recalculating growth and defense strategies under competition: Key roles of photoreceptors and jasmonates. *J. Exp. Bot.*, 70(13), 3425–3434.

Bastías, D.A., Gianoli, E., Gundel, P.E. (2021). Fungal endophytes can eliminate the plant growth–defence trade-off. *New Phytol.*, 230(6), 2105–2113.

Batinic, T., Schmitt, J., Schultz, U.M., Werner, D. (1998). Construction of RAPD-generated DNA probes for the quantification of *Bacillus subtilis* FZB C and the evaluation of its biocontrol efficiency in the system *Cucumis sativus – Pythium ultimum* / Konstruktion von RAPD-Sonden für die Quantifizierung von *Bacillus subtilis* FZB C und dessen antagonistische Wirksamkeit im System *Cucumis sativus – Pythium ultimum*. *J. Plant Dis. Prot.*, 105(2), 168–180.

Bell, J.V., Stewart, A., Rowarth, J.S. (2000). Application method and growing medium affects the response of cucumber seedlings to inoculation with *Trichoderma harzianum*. *Australas. Plant Pathol.*, 29(1), 15–18.

Berkers, E. and Geels, F.W. (2011). System innovation through stepwise reconfiguration: The case of technological transitions in Dutch greenhouse horticulture (1930–1980). *Technol. Anal. Strateg. Manag.*, 23(3), 227–247.

Bodin, E., Bellée, A., Dufour, M.-C., André, O., Corio-Costet, M.-F. (2020). Grapevine stimulation: A multidisciplinary approach to investigate the effects of biostimulants and a plant defense stimulator. *J. Agric. Food Chem.*, 68(51), 15085–15096.

van Butselaar, T. and Van den Ackerveken, G. (2020). Salicylic acid steers the growth–immunity tradeoff. *Trends Plant Sci.*, 25(6), 566–576.

CAN DEPHY (2018). Le glyphosate dans le réseau DEPHY FERME : filière horticulture. November [Online]. Available at: https://ecophytopic.fr/pic/proteger/le-glyphosate-dans-le-reseau-dephy-ferme [Accessed 8 March 2022].

Checker, V.G., Kushwaha, H.R., Kumari, P., Yadav, S. (2018). Role of phytohormones in plant defense: Signaling and cross talk. In *Molecular Aspects of Plant-Pathogen Interaction*, Singh, A. and Singh, I.K. (eds). Springer, Singapore.

van Der Ent, S., Knapp, M., Klapwijk, J., Moerman, E., Van Schelt, J., De Weert, S. (2018). *Knowing & Recognizing: The Biology of Pests, Diseases and their Natural Solutions*. Koppert Biological System, Berkel en Rodenrijs.

Drogue, B., Doré, H., Borland, S., Wisniewski-Dyé, F., Prigent-Combaret, C. (2012). Which specificity in cooperation between phytostimulating rhizobacteria and plants? *Res. Microbiol.*, 163(8), 500–510.

Dubský, M., Šrámek, F., Vosátka, M. (2002). Inoculation of cyclamen, (*Cyclamen persicum*) and poinsettia (*Euphorbia pulcherrima*) with arbuscular mycorrhizal fungi and *Trichoderma harzianum*. *Rostl. Výroba*, 48(2), 63–68.

Dunne, C., Debany, I., Fenton, A., O'Gara, F. (1996). Mechanisms involved in biocontrol by microbial inoculants. *Agronomie*, 16(10), 721–729.

Edmondson, J.L., Cunningham, H., Densley Tingley, D.O. (2020). The hidden potential of urban horticulture. *Nat. Food*, 1(3), 155–159.

Eichmann, R. and Schäfer, P. (2015). Growth versus immunity – A redirection of the cell cycle? *Curr. Opin. Plant Biol.*, 26, 106–112.

Emonet, E. and Rougier, M. (2019). M. DEPHY EXPE, un réseau d'expérimentations pour des systèmes économes en pesticides. *Innovations agronomiques*, 76, 1–2.

Feldmann, F. and Vogler, U. (2021). Towards sustainable performance of urban horticulture: Ten challenging fields of action for modern integrated pest management in cities. *J. Plant Dis. Prot.*, 128(1), 55–66.

FEPEX (2020). Spanish federation of associations of exporting producers of fruit, vegetables, flowers and live plants. The EU produces 166.3 million tonnes of fruit and vegetables, 2.5% less [Online]. Available at: https://www.fepex.es/noticias/detalle/UE-produce-166-millonses-toneladas-frutas-hortalizas [Accessed 8 March 2022].

Figueroa-Macías, J.P., García, Y.C., Núñez, M., Díaz, K., Olea, A.F., Espinoza, L. (2021). Plant growth-defense trade-offs: Molecular processes leading to physiological changes. *Int. J. Mol. Sci.*, 22(2), 693–701.

FranceAgriMer (2019). Chiffres clés de FranceAgriMer. L'horticulture ornementale, données 2017/18 [Online]. Available at: https://www.valhor.fr/fileadmin/A-Valhor/Valhor_PDF/Etudes_Chiffres_cles_FAM_2019.pdf [Accessed 8 March 2022].

Guo, Q., Major, I.T., Howe, G.A. (2018). Resolution of growth-defense conflict: Mechanistic insights from jasmonate signaling. *Curr. Opin. Plant Biol.*, 44, 72–81.

Hamid, B., Zaman, M., Farooq, S., Fatima, S., Sayyed, R.Z., Baba, Z.A., Sheikh, T.A., Reddy, M.S., El Enshasy, H., Gafur, A. et al. (2021). Bacterial plant biostimulants: A sustainable way towards improving growth, productivity, and health of crops. *Sustainability*, 13(5), 1–24.

Harman, G.E. (2006). Overview of mechanisms and uses of *Trichoderma* spp. *Phytopathology*, 96, 190–194.

Hasler, K., Olfs, H.-W., Omta, O., Bröring, S. (2016). Drivers for the adoption of eco-innovations in the German fertilizer supply chain. *Sustainability*, 8(8), 1–18.

Havardi-Burger, N., Mempel, H., Bitsch, V. (2020). Driving forces and characteristics of the value chain of flowering potted plants for the German market. *Eur. J. Hortic. Sci.*, 85(4), 267–278.

Hermosa, R., Rubio, M.B., Cardoza, R.E., Nicolás, C., Monte, E., Gutiérrez, S. (2013). The contribution of Trichoderma to balancing the costs of plant growth and defense. *Int. Microbiol.*, 16(2), 69–80.

Hernández-Rosas, F., Figueroa-Rodríguez, K.A., García-Pacheco, L.A., Velasco-Velasco, J., Sangerman-Jarquín, D.M. (2020). Microorganisms and biological pest control: An analysis based on a bibliometric review. *Agronomy*, 10(11), 1808–1822.

Hilker, M., Schwachtje, J., Baier, M., Balazadeh, S., Bäurle, I., Geiselhardt, S., Hincha, D.K., Kunze, R., Mueller-Roeber, B., Rillig, M.C. et al. (2016). Priming and memory of stress responses in organisms lacking a nervous system. *Biol. Rev. Camb. Philos. Soc.*, 91(4), 1118–1133.

von Hippel, E.A. (1988a). Sources of innovation. SSRN Scholarly Paper ID 2877276. Social Science Research Network, Rochester, New York [Online]. Available at: https://papers.ssrn.com/abstract=2877276 [Accessed 8 March 2022].

von Hippel, E.A. (1988b). *The Sources of Innovation*. Oxford University Press, New York.

Holeski, L.M., Jander, G., Agrawal, A.A. (2012). Transgenerational defense induction and epigenetic inheritance in plants. *Trends Ecol. Evol.*, 27(11), 618–626.

Hulot, J.-F. and Hiller, N. (2021). Exploring the benefits of biocontrol for sustainable agriculture. Institute for European Environmental Policy [Online]. Available at: https://ieep.eu/publications/exploring-the-benefits-of-biocontrol-for-sustainable-agriculture [Accessed 8 March 2022].

IBMA France (2020). Press release of 6 July on Résultats du "Baromètre IBMA France du Biocontrôle" : le biocontrôle poursuit sa progression en 2019 [Online] Available at: https://www.ibmafrance.com/wp-content/uploads/2020/07/200706_CP_Barom%C3%A8tre_IBMA_France_Biocontr%C3%B4le_2019.pdf [Accessed 8 March 2022].

IndustryARC (2020). Biological Control Market – Growth, trends, and forecast (2020–2025) [Online]. Available at: https://www.researchandmarkets.com/reports/3652019/biocontrol-agents-market-forecast-2020-2025 [Accessed 8 March 2021].

Journal officiel de la République Française (2018). Questions orales. Législation européenne relative au biocontrôle, Sénat. no. 82 S., 31 juillet 2018 [Online]. Available at: https://www.senat.fr/seances/s201807/s20180731/s20180731.pdf [Accessed 8 March 2022].

Jullien, J. (2009). Protection biologique des boutures et jeunes plants de Rhododendron contre les maladies fongiques. *PHM-Rev. Hort.*, 512, 20–29.

Kemp, R. and Pearson, P. (2007). Final report MEI project about measuring eco-innovation. *UM Merit, Maastricht*, 32(3), 121–124.

Kesidou, E. and Demirel, P. (2012). On the drivers of eco-innovations: Empirical evidence from the UK. *Res. Policy*, 41(5), 862–870.

Köhl, J., Kolnaar, R., Ravensberg, W.J. (2019). Mode of action of microbial biological control agents against plant diseases: Relevance beyond efficacy. *Front. Plant Sci.*, 10, 845–864.

Lacordaire, A.I., Verdun, C., Piron, M. (2009). Spécial trianum. *BioJournal*, 24, 1–6.

Largant, L. (2019). Les biostimulants en agriculture (et pas seulement) : statut & règlementation. Intervention journées ITAB – Substances naturelles en protection des cultures [Online]. Available at: http://www.afaia.fr/wp-content/uploads/2019/07/afaia-itab-intrants-10-avril-2019.pdf [Accessed 8 March 2021].

López-Moral, A., Agustí-Brisach, C., Trapero, A. (2021). Plant biostimulants: New insights into the biological control of Verticillium wilt of olive. *Front. Plant Sci.*, 12, 662178–662192.

del Mar Tellez, M., Simon, A., Rodriguez, E., Janssen, D. (2017). Control of tomato leaf curl New Delhi virus in zucchini using the predatory mite *Amblyseius swirskii*. *Biol. Control*, 114, 106–113.

Marín-Guirao, J.I., Rodríguez-Romera, P., Lupión-Rodríguez, B., Camacho-Ferre, F., Tello-Marquina, J.C. (2016). Effect of Trichoderma on horticultural seedlings' growth promotion depending on inoculum and substrate type. *J. Appl. Microbiol.*, 121(4), 1095–1102.

Massart, S., Perazzolli, M., Höfte, M., Pertot, I., Jijakli, M.H. (2015). Impact of the omic technologies for understanding the modes of action of biological control agents against plant pathogens. *Biocontrol*, 60(6), 725–746.

Mérillon, J.-M. and Ramawat, K.G. (2020). *Plant Defence: Biological Control.* Springer Nature, Cham.

Mhlongo, M.I., Piater, L.A., Madala, N.E., Labuschagne, N., Dubery, I.A. (2018). The chemistry of plant-microbe interactions in the rhizosphere and the potential for metabolomics to reveal signaling related to defense priming and induced systemic resistance. *Front. Plant Sci.*, 9, 112–129.

Millington Synge, P., Herklots, G.A.C., Perrott, R., Janick, J. (2019). Horticulture. *Encyclopedia Britannica* [Online] Available at: https://www.britannica.com/science/horticulture [Accessed 8 March 2022].

Ministère de l'Agriculture et de l'Alimentation (2021). Note de service DGAL/SDQSPV 2021-200 du 15 mars 2021 [Online]. Available at: https://info.agriculture.gouv.fr/gedei/site/bo-agri/instruction-2021-200 [Accessed 8 March 2022].

Mommaerts, V., Platteau, G., Boulet, J., Sterk, G., Smagghe, G. (2008). Trichoderma-based biological control agents are compatible with the pollinator Bombus terrestris: A laboratory study. *Biol. Control*, 46(3), 463–466.

Mulero-Aparicio, A., Varo, A., Agusti-Brisach, C., López-Escurdo, F.J., Trapero, A. (2020). Biological control of Verticillium wilt of olive in the field. *Crop Prot.*, 128, 104993–105003.

Nicot, P.C. (2011). Classical and augmentative biological control against diseases and pests: Critical status analysis and review of factors influencing their success. IOBC – International Organisation for Biological and Integrated Control of Noxious Animals and Plants.

Octave, S., Amborabé, B.-E., Fleurat-Lessard, P., Bergès, T., Roblin, G. (2006). Modifications of plant cell activities by polypeptides secreted by *Eutypa lata*, a vineyard fungal pathogen. *Physiol. Plant*, 128(1), 103–115.

Oltra, V. and Saint Jean, M. (2005). The dynamics of environmental innovations: Three stylised trajectories of clean technology. *Econ. Innovation New Technol.*, 14(3), 189–212.

O'Regan, R., Wilson, A., Kurtböke, İ. (2019). Use of bacteriophages as biological control agents in horticulture. *Microbiol. Aust.*, 40(1), 47–50.

Pavitt, K. (1984). Sectoral patterns of technical change: Towards a taxonomy and a theory. *Res. Policy*, 13(6), 343–373.

Payne-Gifford, S., Srinivasan, C.S., Dorward, P. (2021). Blunting EU Regulation 1107/2009: Following a regulation into a system of agricultural innovation. *Agric. Human Values*, 38(1), 221–241.

Pétriacq, P., López, A., Luna, E. (2018). Fruit decay to diseases: Can induced resistance and priming help? *Plants*, 7(4), 77–92.

Piasentin, J. (2009). Stimulation des mécanismes de défenses naturelles des plantes – Protection des cultures, production de composés secondaires d'intérêt. Terres d'innovation, France.

Rennings, K. (2000). Redefining innovation – Eco-innovation research and the contribution from ecological economics. *Ecol. Econ.*, 32(2), 319–332.

Rodríguez, E., González, M., Paredes, D., Campos, M., Benítez, E. (2018). Selecting native perennial plants for ecological intensification in Mediterranean greenhouse horticulture. *Bull. Entomol.*, 108(5), 694–704.

Rodriguez, M.V., Tano, J., Ansaldi, N., Carrau, A., Srebot, M.S., Ferreira, V., Martínez, M.L., Cortadi, A.A., Siri, M.I., Orellano, E.G. (2019). Anatomical and biochemical changes induced by *Gluconacetobacter diazotrophicus* stand up for *Arabidopsis thaliana* seedlings from *Ralstonia solanacearum* infection. *Front Plant Sci.*, 10, 1618–1637.

Sinno, M., Ranesi, M., Gioia, L., Errico, G., Lois Woo, S. (2020). Endophytic fungi of tomato and their potential applications for crop improvement. *Agriculture*, 10(12), 587–606.

Šrámek, F., Dubský, M., Vosátka, M. (2000). Effect of arbuscular mycorrhizal fungi and *Trichoderma harzianum* on three species of balcony plants. *Rost Výroba*, 46(3), 127–131.

Stenberg, J.A., Sundh, I., Becher, P.G., Björkman, C., Dubey, M., Egan, P.A., Friberg, H., Gil, J.F., Jensen, D.F., Jonsson, M. et al. (2021). When is it biological control? A framework of definitions, mechanisms, and classifications. *J. Pest Sci.*, 94(3), 665–676.

Tendal, K. and Madsen, A.M. (2011). Exposure to airborne microorganisms, hyphal fragments, and pollen in a field of organically grown strawberries. *Aerobiologia*, 27(1), 13–23.

Valente, J., Gerin, F., Le Gouis, J., Moënne-Loccoz, Y., Prigent-Combaret, C. (2020). Ancient wheat varieties have a higher ability to interact with plant growth-promoting rhizobacteria. *Plant Cell Environ.*, 43(1), 246–260.

Wang, W. and Wang, Z.-Y. (2014). At the intersection of plant growth and immunity. *Cell Host Microbe*, 15(4), 400–402.

Wilson, M. (1997). Biocontrol of aerial plant diseases in agriculture and horticulture: Current approaches and future prospects. *J. Ind. Microbiol.*, 19(3), 188–191.

Yang, B. and Xu, Y. (2021). Applications of deep-learning approaches in horticultural research: A review. *Hortic. Res.*, 8(1), 123–153.

3

Development of Omics Tools for the Assessments of the Environmental Fate and Impact of Biocontrol Agents

Hikmat GHOSSON, Marie-Virginie SALVIA and Cédric BERTRAND
CRIOBE, UAR 3278, University of Perpignan, France

Biocontrol agents and biopesticides are emerging as alternatives to replace synthetic pesticides. This emergence is motivated by the notion that those natural products are less harmful against human health and the environment. Nonetheless, this a priori should be scientifically examined. Thus, risk assessment should be equally exigent for both the synthetic pesticides and the biocontrol agents/biopesticides. However, the lack of adapted and applicable risk assessment protocols for such complex products is an obstacle for their in-depth fate and impact evaluation. Hence, novel approaches and proxies based on omics present promising tools to fulfill this exigence. This chapter focuses on the recent omics-based studies assessing biocontrol agents and biopesticide risks. It introduces novel omics-based proxies aiming to improve (bio)pesticides risk assessment protocols. Furthermore, a succinct glance on future perspectives, requirements and challenges facing those omics approaches will be highlighted.

 © ISTE Ltd 2022.

3.1. Introduction: emergence of biocontrol agents and their risks

The emergence of biocontrol agents (BCA) and biopesticides as an alternative solution to replace synthetic pesticides is motivated by the idea that considers that these emerging products are less harmful against human health and the environment (Copping and Menn 2000; Glare et al. 2012; Villaverde et al. 2014; Pavela and Benelli 2016; Bertrand et al. 2018; Robin and Marchand 2019). Nonetheless, this notion should be scientifically examined. In fact, questions concerning the toxic and ecotoxic risks of biopesticides have already been raised in scientific and regulation communities (Amichot et al. 2018). The given argument behind these questions is that precautions regarding the use of the novel emerging natural substances and their transformation products should be taken into consideration in order to avoid potential risks on human health and the environment. In fact, and as a confirmation of these concerns, recent safety auditing and research works proved that several BCAs and biopesticides could present risks on non-target organisms. For instance, concerns were raised by the European Food Safety Authority (EFSA) regarding the potential toxicity of the insecticide Spinosad (originating from *Saccharopolyspora spinosa*) on human and animal health (European Food Safety Authority et al. 2018). Its ecotoxicity versus *Daphnia magna* model organism (Santos et al. 2019), versus stingless bees *Melipona quadrifasciata* (Tomé et al. 2015) and versus bumble bees *Bombus impatiens* Cresson larval stage in field-like conditions (Morandin et al. 2005) were also reported on. Azadirachtin (an active ingredient of neem oil) was reported to be a potential endocrine disruptor (Robin and Marchand 2019). Direct negative effects of the anti-mosquito bioinsecticide *Bacillus thuringiensis* subsp. *israelensis* on non-target species such as Chlorophyta, Diptera, Lepidoptera and Plecoptera were observed in laboratory experiments (Brühl et al. 2020). In addition, recent preliminary studies investigated a potential impact of the Leptospermone (a natural β-Triketone herbicide) on soil microbial communities (Romdhane et al. 2016, 2019; Mallet et al. 2019).

Hence, the assessment of the risks of BCAs and biopesticides must be considered as an important problem that needs investigating. In an environmental context, the risks can be summarized by two main points: (i) the fate and (ii) the impact of the product.

The environmental fate of a product can be explained by the duration needed for the active substance (AS) to disappear from the environment. The fate concept should also take into consideration the formation of AS's transformation products (TPs), as well as their lifespan in the environment and their degradation. The transformation ways should also ideally be understood. These ways can be either chemical (transformation or degradation by hydrolysis, photolysis, oxidation, reduction, biodegradation, etc.) or physical (leaching, volatilization, bioaccumulation, etc.). The formulation ingredients and their TPs could also be included when assessing the environmental fate of a formulated product.

The assessment of the environmental impact consists of studying the effects of the examined product on non-target living species of the environment, as vertebrate and invertebrate animals, plants or microorganisms. The impact of a BCA or biopesticide can take different forms. It is defined following the assessed factor. For instance, in an agricultural context, the impact can be considered negative when the application of a product causes adverse effects on functional species such as biofertilizers (e.g. *Azospirillum*, *Herbaspirillum*, *Acetobacter*, *Azotobacter* and *Azoarcus* that assure the unavailable nitrogen provision by fixing it from the atmosphere, or the phosphorus solubilizer *Bacillus megaterium*, or the potassium solubilizer *Bacillus mucilaginosus*, etc. (Wu et al. 2005)). As these species play an important role in determining the agricultural quality of the soil, adverse effects on such species might decrease the yield and/or the quality of the agricultural production.

In conclusion, the environmental fate and impact should be deeply investigated and understood in the context of pollution risks assessment. Thus, dedicated experimental designs and analytical protocols should be developed and validated for this purpose. These protocols should be adapted with the complex biochemical nature of BCAs and biopesticides, as well as the biochemical complexity of the environmental matrices.

3.2. Evaluation methodologies: an overview

The classic risk evaluation protocols dedicated to studying synthetic pesticides are so far used to assess BCAs and biopesticide risks. These classic protocols are various and based on different concepts that target different aspects of risks. For instance, fate evaluation is mostly based on studying the persistence and the dissipation of the AS using the DT50 and DT90 concepts.

DT50 is defined as the "disappearance time of 50% of the substance" (DT90 is for 90% of the substance) (European Commission – Directorate General for Agriculture 2000). It targets the known compounds of the AS. It is thus based on targeted analytical approaches with a high selectivity. Concerning the environmental impact assessment, the evaluation concepts are mostly based on the toxicity and the ecotoxicity that are studied using several defined biological models. For instance, the terrestrial species that should be examined are summarized in the Guidance Document on Terrestrial Ecotoxicology Under Council Directive 91/414/EEC (SANCO/10329/2002 rev 2 final) (European Commission – Health & Consumer Protection Directorate-General 2002). They include soil microorganisms. Indeed, the defined indicator in order to assess the ecotoxicity of a Plant Protection Product (PPP) on soil microorganisms is the microbial activity that should not be affected by more than ± 25% after 100 days. It should be mentioned that the toxicity and ecotoxicity tests should be carried out on the active substance (a priori risk assessment) and on the formulation ingredients (a posteriori risk assessment). The formulation ingredients should be tested particularly when a "homologation" of the PPP is required. The "homologation" process is requested when a pesticide AS is suggested for use in order to control different pathogens/target organisms.

3.3. Limitations of classic methodologies

As described previously, the classic methodologies and concepts based on DT50/DT90 only target defined and known chemical compounds, that is, the AS, and in the best cases, the known TPs. From an analytical chemistry point of view, these approaches can present limitations when it comes to BCAs and biopesticides.

In fact, some of those biopesticides are complex mixtures consisting of various known and unknown molecules. These molecules can act in a synergic and/or pleiotropic mode of actions. An example that can be given is the in-course-of-development biofungicide based on *Dittrichia viscosa* extract (Tamm et al. 2017), which is a complex mixture of Terpenoids. Its AS is the "12-Carboxyeudesma-3,11(13)-diene" and its derivatives, but the effective product is the integral plant extract. Another example is the in-course-of-development bioherbicide based on the *Myrica gale* methanolic extract (Popovici et al. 2014). This bioherbicide contains a wide variety of compounds (Triketones, Terpenes and Terpenoids) that were partially identified (Carlton et al. 1992; Svoboda et al. 1998; Popovici et al. 2008;

Ghosson et al. 2020). Among several others, one major active compound is the "Myrigalone A". However, this active compound has been proven to be more efficient when applied within the integral extract (Oracz et al. 2012). This mixture contains several known and unknown polyphenolic compounds that help protect the Myrigalone A from being rapidly degraded by photolysis (Khaled et al. 2019). Hence, as the DT50/DT90 approaches are based on targeted known molecules, they are not applicable when the composition of the PPP is partially unknown. Furthermore, it is difficult to identify and track the TPs that come from such complex mixtures.

On the contrary, the DT50/DT90 approaches are not applicable when it comes to BCAs based on microorganisms. In fact, this type of BCA is not based on the defined molecular AS. The quantification of their active compounds and their TPs is thus not possible. This problematic has been raised during the OECD 9th Biopesticides Expert Group Seminar on Test Methods for Microorganisms (OECD 2019). The report issued following the seminar indicated that the microorganisms-based BCAs are regulated following the same guidelines and approaches that were basically dedicated to chemical pesticides. Thus, it underlines the limitations of such approaches for this type of BCA, as their nature and mechanisms of actions are different. The report also considers these approaches to be outdated, as recent scientific advances have carried out novel techniques and proxies that can be exploited for microorganisms-based BCAs risk assessment. Therefore, new alternatives are recommended for this task.

Besides, it is also important to mention that the classic indicators used in order to assess the ecotoxicity/impact of pesticides on soil microorganisms – based on measuring the microbial activity – have their limitations. In fact, the microbial activity is assessed by measuring the nitrogen and carbon mineralization processes (Council of the European Union 1997; OECD 2000a, 2000b). This approach allows one to acquire global information about the impact, but no specific indications can be given about the impacted microbial species and strains, nor about effects on soil biodiversity and the functional microbial communities (Martin-Laurent et al. 2013; Thiour-Mauprivez et al. 2019). Hence, the classic concept for ecotoxicity assessment is not able to determine the impacted microorganisms or if the impact can be linked to a potential deterioration of soil quality (i.e. if the functional species are negatively impacted). No information can thus be given by those protocols for phytopharmacovigilance or potential bioremediation purposes. This issue was underlined by the EFSA, as stated in their scientific opinion published in 2010 (EFSA Panel on Plant Protection Products and their Residues (PPR) 2010). In this paper, they point out the importance of the functional groups when defining the key drivers of impacts on relevant ecosystem services. The defined key drivers are "the major taxonomic or functional groups that provide the ecosystem service". In 2017, the EFSA published another scientific opinion that widely described and detailed the state of the art of risk assessment of PPPs for in-soil organisms (EFSA Panel on Plant Protection Products and their Residues (PPR) et al. 2017). The paper also highlighted the specific protection goals for those organisms including mycorrhiza, fungi, protozoa, bacteria and archaea, among others.

3.4. Omics: potential tools for risks assessment?

As the classic protocols for risk assessment suffer from fundamental and various limitations when applied to BCAs and biopesticides, the evaluation of those products using classic guidelines seems to be non-optimal. Thus, advanced approaches should be set up. Those approaches should be able to deal with the main obstacles limiting the performance of the classic protocols for BCAs and biopesticides risk assessment. For instance, the novel approaches that must be developed in order to study the environmental fate should be capable of analyzing both the known and the unknown molecules composing the AS, as well as their TPs. The formulation ingredients should ideally be analyzed as well. On the contrary, the impact assessment should take into consideration the affected microorganism species, strains and their functions. It should also allow one to understand the nature of the impact depending on the context of the study (e.g. agricultural, ecological). This consideration can provide some essential basis that can be used to enhance decision-making procedures, leading to safe and sustainable agricultural activities, ecosystems protection and possible bioremediation processes when it is needed.

In this context, omics- and meta-omics-based approaches seem to be exploitable. The omics discipline can be divided into four main sub-disciplines: genomics, transcriptomics, proteomics and metabolomics. Each of these sub-disciplines aim to study a different biomolecular scale: genomics for DNA, transcriptomics for RNA, proteomics for the proteins and metabolomics for the metabolites. The objective of omics as described by M. Vailati-Riboni and colleagues "is to identify, characterize, and quantify all biological molecules that are involved in the structure, function, and dynamics of a cell, tissue, or organism" (Vailati-Riboni et al. 2017). The meta-omics are omics applied to a community scale. For instance, the soil meta-metabolomics aims to study the metabolome originating from the whole microbial community of a given soil (Dubitzky et al. 2013; O'Malley 2013). Omics and meta-omics are known to be holistic approaches that can collect an important amount of information when studying biological models and matrices. Therefore, these disciplines can afford potential means for developing novel concepts and proxies for (bio)pesticides and BCAs risks assessment. They recently started to be used for studying synthetic pesticides biodegradation. These works are reported and detailed in the review by A. Rodríguez and colleagues (2020). This chapter will focus on genomics- and metabolomics-based studies for pesticides, biopesticides and BCAs risks assessment, as most of the reported research works mainly lay on those two sub-disciplines. Only a few other works reported proteomics-based studies (Seo et al. 2013; Tiwari et al. 2018), as well as transcriptomics-based studies that were carried out to assess the impact of different synthetic pesticides on various microbial species and strains (Dennis et al. 2003; Jang et al. 2008; Nde et al. 2008; Allen and Griffiths 2012; Lu et al. 2013; Li et al. 2015; Namouchi et al. 2016; Gil et al. 2018; Meng et al. 2019; Mesnage et al. 2020).

3.4.1. *Genomics for (bio)pesticides and BCAs risks assessment*

Genomics- and meta-genomics-based approaches allow one to screen and quantify living (micro)organisms in an environment at a given point in time. This can be achieved by holistic or specific sequencings of the sample's DNA using different molecular biology techniques such as PCR, qPCR, T-RFLP, DGGE and others, which allow one to detect and quantify the different species and/or strains that are present in the analyzed sample. Those approaches are being increasingly used for the assessment of synthetic pesticides, biopesticides and BCAs' effects on microorganisms in the environment. Several genomics-based research works studying the impact of BCAs and (bio)pesticides on environmental microbial communities were reported in the literature. They are summarized in Table 3.1.

Study	BCA/biopesticide	Synthetic pesticide	Sample	Experimental conditions	Used methods and techniques
(Savazzini et al. 2009)	*Trichoderma atroviride* SC1	None	Soil	Field trial	Real-time PCR, ARISA
(Cordier and Alabouvette 2009)	*Trichoderma strain* I-1237	None	Two different soils	Growth chamber-controlled soil mesocosms	T-RFLP, PCR
(Correa et al. 2009)	*Bacillus amyloliquefaciens* BNM122	Thiram and Carbendazim mix	Soil and rhizosphere soil	Greenhouse experiment	PCR, DGGE
(Kröber et al. 2014)	*Bacillus amyloliquefaciens* FZB42	None	Rhizosphere soil	Field trial	Whole metagenome sequencing
(Li et al. 2016)	*Bacillus subtilis* B068150	None	Three different rhizosphere and non-rhizosphere soils	Pot experiment	PCR, DGGE
(Chen et al. 2018)	*Bacillus subtilis* NCD-2	Dazomet	Rhizosphere soil	Greenhouse experiment	Real-time PCR, DGGE
(Schreiter et al. 2018)	*Pseudomonas* sp. RU47	None	Three different soils	Field trial	DGGE
(Fournier et al. 2020)	*Clonostachys rosea* f. *catenulata*	Fosetyl-Al and Propamocarb-hydrochloride mix	Loamy soil	Controlled soil mesocosms	PCR, metabarcoding
(Romdhane et al. 2016)	Leptospermone	None	Two different soils	Soil microcosms	qPCR, pyrosequencing of 16S rRNA gene amplicons
(Mallet et al. 2019)	Leptospermone	None	Two different soils	Soil microcosms	Sequencing of the internal transcribed spacer region of the fungal rRNA

PCR: Polymerase Chain Reaction; ARISA: Automated Ribosomal Intergenic Spacer Analysis; T-RFLP: Terminal Restriction Fragment Length Polymorphism; DGGE: Denaturing Gradient Gel Electrophoresis; qPCR: Quantitative Polymerase Chain Reaction; rRNA: Ribosomal Ribonucleic Acid.

Table 3.1. *Summary of the reported genomics-based research works conducted in order to assess the impact of BCAs and (bio)pesticides on microbial communities*

These reported works were majorly carried out in order to study the environmental impact of microorganisms-based anti-fungus BCAs. They were applied on soil and/or rhizosphere in various experimental conditions. For instance, Savazzini and colleagues studied the impact of *Trichoderma atroviride* SC1 anti-fungus BCA on soil microbial communities of vineyards in a field trial (Savazzini et al. 2009). The work concluded that the introduction of the BCA had an effect on the soil microflora during the first two weeks following inoculation. However, later, the influence of the environmental conditions outweighed the effects of BCA application. Cordier and Alabouvette assessed the effects of the introduction of the *Trichoderma atroviride* strain I-1237 (anti-fungus BCA) on non-target soil microorganisms in a growth chamber-controlled soil mesocosms (Cordier and Alabouvette 2009). Two soils with different physicochemical properties were used and the study concluded that the introduction of a BCA into two soils slightly modified the microbial diversity for a short period, and the resilience took place nine months after BCA application.

Other works were carried out in order to study Bacillus-based BCAs. Correa and collaborators assessed the impact of *Bacillus amyloliquefaciens* BNM122 anti-fungus BCA on rhizosphere and soil microbial communities (Correa et al. 2009). The experiments were achieved by treating soybean seeds in a greenhouse. A comparison with a chemical fungicide based on a mix of Thiram and Carbendazim was also performed. The work concluded that neither the ecophysiological structure nor the physiological profiles of the soybean rhizosphere bacterial community showed important changes after treatment with the BCA. However, the treatment with chemical fungicides altered the metabolic profiles of the cultivable bacterial community of the rhizosphere. On the contrary, Kröber and co-workers studied the effect of the strain *Bacillus amyloliquefaciens* FZB42 (anti-fungus BCA) on the microbial community in the lettuce rhizosphere (Kröber et al. 2014). Whole Metagenome Sequencing was performed on rhizosphere soil in a field trial. Marginal changes were revealed after the application of the BCA, and no major effects were observed at the taxonomic and functional scales. Besides, Li and collaborators investigated the response to *Bacillus subtilis* B068150 anti-fungus BCA by the microbial community of cucumber rhizosphere (Li et al. 2016). The study was carried out on three different rhizosphere and non-rhizosphere soils in a pot experiment. Results showed that the BCA introduction did not significantly change the diversity of rhizosphere microbial communities of three soils. However, a soil-specific impact was observed with a significant negative correlation between population density of the BCA strain and the bacterial diversity in the clay soil rhizosphere, which was not significant in loam and sandy soils. Another recent study on Bacillus-based BCAs was achieved by Chen and co-workers. The work evaluated the effects of *Bacillus subtilis* NCD-2 (anti-fungus BCA) and Dazomet (chemical fungicide) on rhizosphere soil microbiota

(Chen et al. 2018). A series of two-year greenhouse experiments allowed one to conclude that chemical treatment significantly decreased the soil catalase and urease activities compared with the control, while the BCA application showed a significant increase in bacterium/fungus ratios, soil urease and acid phosphatase activities.

Besides, studies on other microorganisms-based anti-fungus BCA were also reported, such as *Pseudomonas* sp. RU47 (Schreiter et al. 2018) and *Clonostachys rosea* f. *catenulata* (Fournier et al. 2020). The first study by Schreiter et al. assessed the impact of the BCA on bacterial communities of three different potato and lettuce rhizosphere soils in a field trial. Results showed that the effect of BCA application on bacterial community composition was stronger in the rhizosphere of lettuce than in the potato rhizosphere, and that a significant effect of the BCA on the *Pseudomonas*-specific *gacA* fingerprints of the rhizosphere was only observed in lettuce in alluvial soil. The work concluded that the BCA influence on the rhizosphere bacterial community could be an important criterion for its registration and use (Schreiter et al. 2018). The second work by Fournier et al. investigated the impact of the BCA in comparison with a synthetic fungicide (based on Fosetyl-Al and Propamocarb-hydrochloride mix) on soil bacterial, fungal and protist communities. This experiment was conducted on a loamy soil in controlled soil mesocosms conditions. Results showed that the BCA and the synthetic pesticide have different impact on soil functioning, and that pest management products may have functionally significant impacts on the soil microbiome. The study concluded that high-throughput sequencing analyses might be a promising tool for the ecotoxicological risk assessment of pesticides and BCAs on non-target soil microorganisms (Fournier et al. 2020).

It is worth mentioning that genomics-based evaluation of the ecotoxic impact of a bioherbicide, the Leptospermone on soil microbial community and soil fungal composition and diversity, were reported in the literature (Romdhane et al. 2016; Mallet et al. 2019). The first study by Romdhane et al. focused on microbial communities of two different arable soils. A microcosm-scale experiment was conducted through a 45-day kinetic tracking. 11 microbial groups were targeted by qPCR and pyrosequencing of 16S rRNA gene amplicons after the application of one-time and ten-times the bioherbicide recommended dose. Results were compared to a screening preformed on untreated control microcosms and showed shifts in the bacterial community structure and a significant impact of Leptospermone on the diversity of the soil bacterial communities of the two soils. The resilience of those communities was observed only when the Leptospermone fully disappeared (Romdhane et al. 2016). The second study by Mallet et al. focused on the fungal composition and diversity in soil. The same experimental design by Romdhane et al. was applied (the application of ten-times the recommended dose was excluded from the protocol) and a sequencing of the internal transcribed spacer region of the fungal rRNA was performed. Results revealed significant changes in the fungal community structure and diversity in both soils, with an incomplete recovery of the fungal community in one of the two soils even after the dissipation of the bioherbicide (Mallet et al. 2019).

3.4.2. *Metabolomics for (bio)pesticides and BCAs risks assessment*

Metabolomics is an analytical chemistry discipline that considers studying "small biomolecules" with molecular mass lower than 1,000–1,500 Da. This discipline has been increasingly growing and advancing for more than two decades (according to PubMed[1], the first paper holding the term "metabolome" was in 1998 by H. Tweeddale and colleagues (1998)). Metabolomics has been developed and applied in a wide range of fields that includes medicine, pharmacology, toxicology, forensics, nutrition and food security, livestock, marine biology, plant biology, agriculture and environment. Its approaches, and particularly the untargeted metabolomics, have been proven capable for high throughput analyses that provide a big amount of data, using performant analytical chemistry techniques, relatively short-time analyses and computational data processing tools.

The involvement of metabolomics in the environmental research is increasingly considered in recent years (Lin et al. 2006; Bundy et al. 2009; Lankadurai et al. 2013; Viant and Sommer 2013; Longnecker et al. 2015; Bedia et al. 2018). The main aims of environmental metabolomics are to study the abiotic stress, the anthropogenic influence and the impact of environmental diseases on nature and its existing species (Bundy et al. 2009). Environmental biomarkers and the modifications in the metabolic pathways are searched following factors such as temperature or acidity change in the studied environment, or following the pollution of the environment by chemicals or its invasion by intruder species. Nonetheless, the environmental metabolomics is still a "minor" interest if compared to the

1. Pubmed.ncbi.nlm.nih.gov/?term=(metabolomics), accessed June 23, 2021.

global state of metabolomics research. Figure 3.1 shows that between 2010 and 2021, papers holding both terms "environmental" and "metabolomics"[2] did not surpass 26.00% of the total number of papers holding the term "metabolomics"[3] (the maximum reached was 25.81% in 2019). In fact, environmental metabolomics still confront several challenging problematics as sample designing, extraction protocols, data processing and the significant lack of environmental metabolome databases (Bundy et al. 2009).

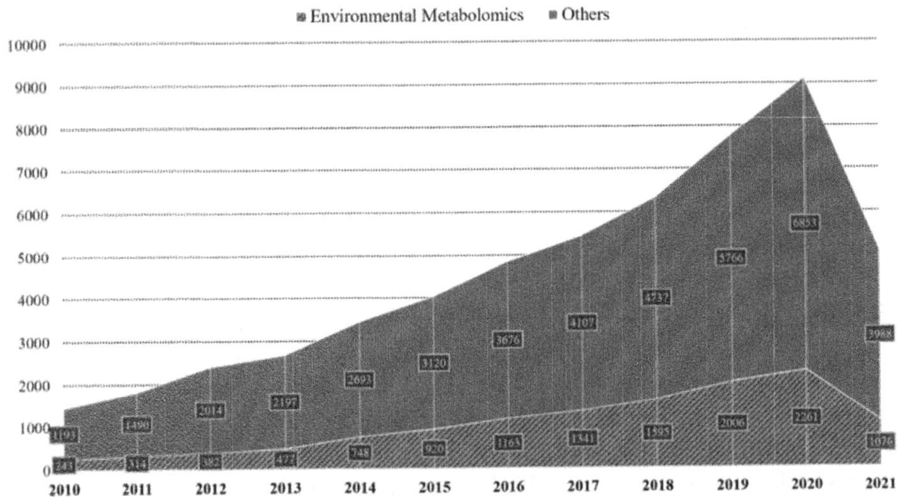

Figure 3.1. *Evolution of the number of articles holding the term "metabolomics", or both terms "environmental" and "metabolomics" between 2010 and 2021, according to the PubMed database (June 23, 2021) ("Search for the term 'Metabolomics'", 2021; "Search for the terms 'Metabolomics' and 'Environmental'", 2021). For a color version of this figure, see www.iste.co.uk/prigentcombaret/biocontrol.zip*

Untargeted metabolomics applied for pesticides research was also reported (Aliferis and Chrysayi-Tokousbalides 2011). It mostly aims to elucidate the modes-of-action of bioactive compounds, ecotoxicity and toxicity assessment, and to investigate pesticides effects on plant metabolic pathways. In addition, metabolomics for studying synthetic pesticides biodegradation was reported in the literature (Rodríguez et al. 2020). Several of the reported studies were conducted to determine the biodegradation

2. Pubmed.ncbi.nlm.nih.gov/?term=(metabolomics)%20AND%20(environmental), accessed June 23, 2021.

3. Pubmed.ncbi.nlm.nih.gov/?term=(metabolomics), accessed June 23, 2021.

pathways of pesticides' ASs by different microorganisms and to identify their degradation by-products (Lü et al. 2003; Lénárt et al. 2013; Chen et al. 2015; Wolfand et al. 2016; Meng et al. 2019; Li et al. 2020). The objective of those works was to discover new potential toxic pesticides TPs and/or to seek bioremediation possibilities using microbial species and strains that are capable of biodegrading synthetic pesticides. Other reported studies aimed to investigate ASs' effects and toxicity on microbial communities by studying a microbial metabolome (Bhat et al. 2015; Szewczyk et al. 2015), lipidome (Bernat et al. 2018) or volatolome (Hidalgo et al. 2019). Or, they aimed to determine both the biodegradation and the impact of ASs on different types of microorganisms (Seo et al. 2013; Feng et al. 2017). All those studies were however performed using microbial growth media. No soil microcosm or field experiments have been reported so far. Only one lipidomics-based study was conducted in a field trial. It investigated the persistence in soil of anti-fungus BCAs (*Bacillus subtilis* GB03, *Burkholderia ambifaria*-type Wisconsin isolate J82, *Trichoderma virens* Gl-21 and *Trichoderma harzianum* strain T-22), as well as the chemical fungicide Thiophanate-methyl/Mancozeb/Cymoxanil (TMC) mixture. The study also evaluated the impact of the different BCAs and pesticide application on soil microbial communities, as well as the impact of the application of a combined BCA/chemical treatment (Larkin 2016).

Besides, as the untargeted metabolomics seems to be a promising tool that can provide high throughput generation of biochemical information, it was recently considered in order to suggest a novel "universal" tool for assessing both the environmental fate and impact of complex (bio)pesticides in the environmental matrices. The suggested approach was called "Environmental Metabolic Footprinting" (EMF), introduced in 2016–2017 by Patil et al. (2016) and Salvia et al. (2018). As shown in Figure 3.2, the EMF is based on a meta-metabolomics approach (i.e. metabolomics of a whole community (Dubitzky et al. 2013; O'Malley 2013)) that aims to extract, analyze and detect both the xenometabolome of an applied pesticide, and the endometabolome of the environmental matrix. The xenometabolome consists of the AS, the formulation ingredients and the TPs issued either from the AS or from the formulation ingredients. The endometabolome consists of metabolites produced by the different microorganisms living in the studied environmental matrix, as the primary metabolites and the secondary metabolites. The xenometabolome and the endometabolome will then constitute the meta-metabolome that will be the target of the extraction, the chemical analyses and the data processing.

The EMF seeks to investigate integrally the persistence, the transformation and the impact of an applied product to an environmental matrix. As an untargeted approach, it can consider either known compounds or unknown compounds. Thus, it must be applicable for both the synthetic pesticides and the biopesticides. In addition, as it aims to analyze the whole meta-metabolome, the fate and the impact of the complex formulated products or microorganisms-based products can be studied. The EMF should also detect new TPs, and determine new biomarkers of impact. This can be done by comparative kinetics experiments, where the polluted (treated) matrix is compared to an unpolluted control matrix.

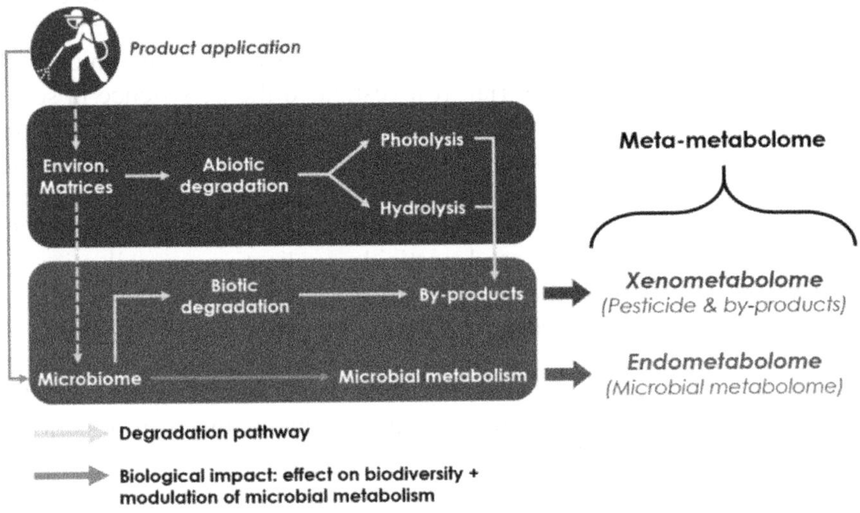

Figure 3.2. *The environmental metabolic footprinting concept. Figure readapted from Patil et al. (2016), with permission from the authors. For a color version of this figure, see www.iste.co.uk/prigentcombaret/biocontrol.zip*

The ultimate aim of EMF is to introduce a novel indicator called the "resilience time". This indicator is defined as the time needed for the difference between the meta-metabolome of the treated matrix and that of the control matrix to be statistically non-significant (Figure 3.3). In this case, the biochemical state of the environmental matrix and its community is considered to be "normalized", as it is comparable to the control matrix, that is, the "natural" state of the matrix. Thus, at the resilience time, the whole xenometabolome including all pollutants issued from the pesticide are considered to be dissipated (or more precisely, below the detection limit of the approach), and the observable impact of the pollution is surpassed.

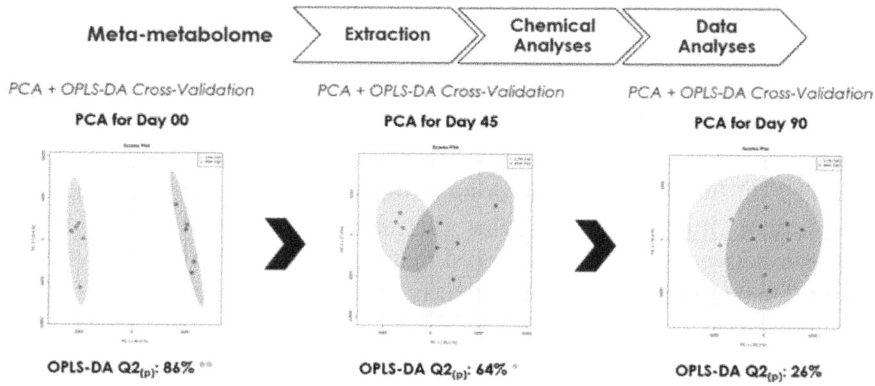

Figure 3.3. *The "resilience time" indicator. This example shows that the resilience time is reached between day 45 and day 90, as the predictivity of the "p" component (Q2(p)) of the OPLS-DA model decreased from 64% to 26% between those two time points (discrimination is considered insignificant in metabolomics if Q2(p) is lower than 50%). For a color version of this figure, see www.iste.co.uk/prigentcombaret/biocontrol.zip*

The previously mentioned research works that introduced the EMF aimed to prove the concept of the approach. The first by Patil et al. (2016) was carried out on soil microcosms. It investigated the fate and the impact of two pure β-Triketone herbicides with a 90-day kinetics: the Sulcotrione as a synthetic product and the Leptospermone as its natural alternative. This work was followed by another study that combined the meta-metabolomics-based EMF approach with the next-generation sequencing of 16S rRNA gene amplicons (metabarcoding). Metabarcoding was included in order to investigate the impact of the studied pesticides on the diversity and activity of the soil bacterial community (Romdhane et al. 2019). This study found that Leptospermone caused stronger modifications of soil bacterial diversity than the Sulcotrione (at ten-times the field dose for both), and that no resilience was observed even after the complete dissipation of Leptospermone. It also proved the advantage of combining metabolomics with the 16S rRNA sequencing, by correlating the ecotoxic effects on the microbial community to the appearance of the TPs of Leptospermone and Sulcotrione, such as the Hydroxy-Leptospermone and the 2-Chloro-4-Methylsulfonyl Benzoic Acid, respectively. The impact on the soil microbial community was also correlated to biomarkers belonging to soil endometabolome. However, further studies must be carried out in order to find a relation between these data (endometabolites–soil microbial community relation).

The second work about the EMF by Salvia et al. (2018) proved the applicability of the EMF on a complex microorganism-based bioinsecticide: the *Bacillus thuringiensis israelensis* (Bti). The proof-of-concept study was performed on two kinetics time points after the application of the pesticide on sediment microcosm samples. It showed a significant statistical difference between control and spiked meta-metabolomes after 8 days of the spiking. This approach seems sufficiently sensitive for such microorganisms-based pesticides.

On the contrary, the untargeted metabolomics approach to study the volatile residues of (bio)pesticides was recently developed and proven to be applicable in order to investigate the fate of complex plant extracts pesticides (Ghosson et al. 2020). It was applied to study a typical bioherbicide; the *Myrica gale* extract applied on soil. The study was achieved in a laboratory soil microcosm experiment that went through a 38-day kinetics study. It showed the capabilities of untargeted metabolomics tools (i.e. the high throughput HS-SPME–GC–MS analyses, the data preprocessing and the statistical analyses) to track the evolution of the residues profile over the time, and to determine and identify the complex composition of plant extracts-based pesticides and their TPs.

3.5. Perspectives

According to the literature and to the actual state of the art, omics-based approaches seem to be significantly promising for the assessment of BCAs and complex (bio)pesticides environmental fate and impact. However, the development stage of those approaches does not allow us to fulfill the expected ambitions and long-term perspectives. Indeed, several important challenges should be tackled, and in-depth evaluations and improvements of omics are required for the fundamental and technical scales.

For instance, metabolomics is a powerful approach that seems to be capable of the collection and analysis of a high amount of biochemical information. However, this approach still suffers from several drawbacks, such as the critical step of sample preparation, chemical extraction, physicochemical analyses and data preprocessing. Those factors and their parameters are experiment- and matrix-dependent. They are also vulnerable to experimental design and/or matrix complexity and matrix effects (Wu and Li 2016; Ghosson et al. 2021) and should be compromised in order to allow the collection of the interesting and/or the maximum possible amount of data (Ghosson 2020).

The multi-platforms metabolomics approaches (e.g. LC–MS, GC–MS, UHRMS, NMR) as well as multi-omics studies present a wide and interesting field of scientific investments. This type of approach allows one to widen the range of the collectible biological–chemical information, and thus provides highly confident and robust results and conclusions. Nonetheless, these multi-approaches require massive investments on the technological scale, as well as deep developments in the computational and statistical analyses, which presents an important obstacle to any ambitions for further improvements of such cutting-edge research.

3.6. List of abbreviations

ARISA	Automated Ribosomal Intergenic Spacer Analysis
AS	Active Substance
BCA	Biocontrol Agent
DGGE	Denaturing Gradient Gel Electrophoresis
DNA	Deoxyribonucleic Acid
EFSA	European Food Safety Authority
EMF	Environmental Metabolic Footprinting
GC–MS	Gas Chromatography–Mass Spectrometry
HS-SPME–GC–MS	Headspace-Solid Phase Microextraction–Gas Chromatography–Mass Spectrometry

LC–MS	Liquid Chromatography–Mass Spectrometry
mRNA	Messenger Ribonucleic Acid
NMR	Nuclear Magnetic Resonance
OECD	Organization for Economic Co-operation and Development
PCR	Polymerase Chain Reaction
PPP	Plant Protection Product
qPCR	Quantitative Polymerase Chain Reaction
RNA	Ribonucleic Acid
RNA-Seq	Ribonucleic Acid Sequencing
rRNA	Ribosomal Ribonucleic Acid
T-RFLP	Terminal Restriction Fragment Length Polymorphism
TP	Transformation Product
UHRMS	Ultra-High Resolution Mass Spectrometry

3.7. Acknowledgments

The authors acknowledge Chandrashekhar Patil, PhD (University of Illinois at Chicago) for the interesting scientific discussions that allowed the enhancement of the manuscript.

The current work is a part of the PhD thesis of Hikmat Ghosson, PhD (NNT: 2020PERP0029). The PhD grant was awarded to him by the French Ministry of Higher Education, Research and Innovation (MESRI), via the Doctoral School "Energie et Environnement" (ED 305) – University of Perpignan. The fellowship research was co-funded by the French National Research Agency (ANR) under the TRICETOX project (ANR-13-CESA-0002), the European Regional Development Fund (ERDF) under the Interreg POCTEFA PALVIP project (POCTEFA 2014-2020) and the Marie

Skłodowska-Curie Actions EnvFate project (Grant agreement ID: 746656, funded under H2020-EU.1.3.2).

3.8. References

Aliferis, K.A. and Chrysayi-Tokousbalides, M. (2011). Metabolomics in pesticide research and development: Review and future perspectives. *Metabolomics*, 7(1), 35–53.

Allen, K.J. and Griffiths, M.W. (2012). Impact of hydroxyl- and superoxide anion-based oxidative stress on logarithmic and stationary phase *Escherichia coli* O157:H7 stress and virulence gene expression. *Food Microbiology*, 29(1), 141–147.

Amichot, M., Joly, P., Martin-Laurent, F., Siaussat, D., Lavoir, A.-V. (2018). Biocontrol, new questions for ecotoxicology? *Environmental Science and Pollution Research*, 25(34), 33895–33900.

Bedia, C., Cardoso, P., Dalmau, N., Garreta-Lara, E., Gómez-Canela, C., Gorrochategui, E., Navarro-Reig, M., Ortiz-Villanueva, E., Puig-Castellví, F., Tauler, R. (2018). Chapter 19 – Applications of metabolomics analysis in environmental research. In *Comprehensive Analytical Chemistry*, Jaumot, J., Bedia, C., Tauler, R. (eds). Elsevier, Amsterdam.

Bernat, P., Nykiel-Szymańska, J., Stolarek, P., Słaba, M., Szewczyk, R., Różalska, S. (2018). 2,4-Dichlorophenoxyacetic acid-induced oxidative stress: Metabolome and membrane modifications in *Umbelopsis isabellina*, a herbicide degrader. *PLoS One*, 13(6), e0199677.

Bertrand, C., Prigent-Combaret, C., Gonzales-Coloma, A. (2018). Chemistry, activity, and impact of plant biocontrol products. *Environmental Science and Pollution Research*, 25(30), 29773–29774.

Bhat, S.V., Booth, S.C., Vantomme, E.A.N., Afroj, S., Yost, C.K., Dahms, T.E.S. (2015). Oxidative stress and metabolic perturbations in *Escherichia coli* exposed to sublethal levels of 2,4-dichlorophenoxyacetic acid. *Chemosphere*, 135, 453–461.

Brühl, C.A., Després, L., Frör, O., Patil, C.D., Poulin, B., Tetreau, G., Allgeier, S. (2020). Environmental and socioeconomic effects of mosquito control in Europe using the biocide *Bacillus thuringiensis* subsp. *israelensis* (Bti). *Science of the Total Environment*, 724, 137800.

Bundy, J.G., Davey, M.P., Viant, M.R. (2009). Environmental metabolomics: A critical review and future perspectives. *Metabolomics*, 5(1), 3–21.

Carlton, R.R., Waterman, P.G., Gray, A.I. (1992). Variation of leaf gland volatile oil within a population of sweet gale (*Myrica gale*) (Myricaceae). *Chemoecology*, 3(1), 45–54.

Chen, S., Deng, Y., Chang, C., Lee, J., Cheng, Y., Cui, Z., Zhou, J., He, F., Hu, M., Zhang, L.-H. (2015). Pathway and kinetics of cyhalothrin biodegradation by *Bacillus thuringiensis* strain ZS-19. *Scientific Reports*, 5(1), 8784.

Chen, H., Zhao, S., Zhang, K., Zhao, J., Jiang, J., Chen, F., Fang, W. (2018). Evaluation of soil-applied chemical fungicide and biofungicide for control of the fusarium wilt of chrysanthemum and their effects on rhizosphere soil microbiota. *Agriculture*, 8(12), 184.

Copping, L.G. and Menn, J.J. (2000). Biopesticides: A review of their action, applications and efficacy. *Pest Management Science*, 56(8), 651–676.

Cordier, C. and Alabouvette, C. (2009). Effects of the introduction of a biocontrol strain of *Trichoderma atroviride* on non target soil micro-organisms. *European Journal of Soil Biology*, 45(3), 267–274.

Correa, O.S., Montecchia, M.S., Berti, M.F., Fernández Ferrari, M.C., Pucheu, N.L., Kerber, N.L., García, A.F. (2009). *Bacillus amyloliquefaciens* BNM122, a potential microbial biocontrol agent applied on soybean seeds, causes a minor impact on rhizosphere and soil microbial communities. *Applied Soil Ecology*, 41(2), 185–194.

Council of the European Union (1997). Council Directive 97/57/EC: Establishing Annex VI to Directive 91/414/EEC concerning the placing of plant protection products on the market. Council Directive, 97/57/EC, 23.

Dennis, P., Edwards, E.A., Liss, S.N., Fulthorpe, R. (2003). Monitoring gene expression in mixed microbial communities by using DNA microarrays. *Applied and Environmental Microbiology*, 69(2), 769–778.

Dubitzky, W., Wolkenhauer, O., Yokota, H., Cho, K.-H. (eds) (2013). *Encyclopedia of Systems Biology*, 1st edition. Springer-Verlag, New York.

EFSA Panel on Plant Protection Products and their Residues (PPR) (2010). Scientific Opinion on the development of specific protection goal options for environmental risk assessment of pesticides, in particular in relation to the revision of the Guidance Documents on Aquatic and Terrestrial Ecotoxicology (SANCO/3268/2001 and SANCO/10329/2002). *EFSA Journal*, 8(10), 1821.

EFSA Panel on Plant Protection Products and their Residues (PPR), Ockleford, C., Adriaanse, P., Berny, P., Brock, T., Duquesne, S., Grilli, S., Hernandez-Jerez, A.F., Bennekou, S.H., Klein, M. et al. (2017). Scientific Opinion addressing the state of the science on risk assessment of plant protection products for in-soil organisms. *EFSA Journal*, 15(2), e04690.

European Commission – Directorate General for Agriculture (2000). Guidance Document on Persistence in Soil (9188/VI/97 rev. 8). European Commission.

European Commission – Health & Consumer Protection Directorate-General (2002). Guidance document on terrestrial ecotoxicology under Council Directive 91/414/EEC (SANCO/10329/2002 rev 2 final). European Commission.

European Food Safety Authority, Arena, M., Auteri, D., Barmaz, S., Brancato, A., Brocca, D., Bura, L., Carrasco Cabrera, L., Chiusolo, A., Court Marques, D. et al. (2018). Peer review of the pesticide risk assessment of the active substance spinosad. *EFSA Journal*, 16(5), e05252.

Feng, W., Wei, Z., Song, J., Qin, Q., Yu, K., Li, G., Zhang, J., Wu, W., Yan, Y. (2017). Hydrolysis of nicosulfuron under acidic environment caused by oxalate secretion of a novel Penicillium oxalicum strain YC-WM1. *Scientific Reports*, 7(1), 647.

Fournier, B., Pereira Dos Santos, S., Gustavsen, J.A., Imfeld, G., Lamy, F., Mitchell, E.A.D., Mota, M., Noll, D., Planchamp, C., Heger, T.J. (2020). Impact of a synthetic fungicide (fosetyl-Al and propamocarb-hydrochloride) and a biopesticide (*Clonostachys rosea*) on soil bacterial, fungal, and protist communities. *Science of the Total Environment*, 738, 139635.

Ghosson, H. (2020). Development of a novel universal proxy to assess the environmental fate and impact of complex (bio)pesticides by mass spectrometry-based metabolomics. Doctoral Thesis, University of Perpignan Via Domitia, Perpignan.

Ghosson, H., Raviglione, D., Salvia, M.-V., Bertrand, C. (2020). Online headspace-solid phase microextraction-gas chromatography-mass spectrometry-based untargeted volatile metabolomics for studying emerging complex biopesticides: A proof of concept. *Analytica Chimica Acta*, 1134, 58–74.

Ghosson, H., Guitton, Y., Ben Jrad, A., Patil, C., Raviglione, D., Salvia, M.-V., Bertrand, C. (2021). Electrospray ionization and heterogeneous matrix effects in liquid chromatography/mass spectrometry based meta-metabolomics: A biomarker or a suppressed ion? *Rapid Communications in Mass Spectrometry*, 35(2), e8977.

Gil, F.N., Gonçalves, A.C., Becker, J.D., Viegas, C.A. (2018). Comparative analysis of transcriptomic responses to sub-lethal levels of six environmentally relevant pesticides in Saccharomyces cerevisiae. *Ecotoxicology*, 27(7), 871–889.

Glare, T., Caradus, J., Gelernter, W., Jackson, T., Keyhani, N., Köhl, J., Marrone, P., Morin, L., Stewart, A. (2012). Have biopesticides come of age? *Trends in Biotechnology*, 30(5), 250–258.

Hidalgo, K., Ratel, J., Mercier, F., Gauriat, B., Bouchard, P., Engel, E. (2019). Volatolomics in bacterial ecotoxicology, a novel method for detecting signatures of pesticide exposure? *Frontiers in Microbiology*, 9, 3113.

Jang, H.-J., Nde, C., Toghrol, F., Bentley, W.E. (2008). Microarray analysis of toxicogenomic effects of ortho-phenylphenol in *Staphylococcus aureus*. *BMC Genomics*, 9(1), 411.

Khaled, A., Sleiman, M., Darras, E., Trivella, A., Bertrand, C., Inguimbert, N., Goupil, P., Richard, C. (2019). Photodegradation of Myrigalone A, an Allelochemical from *Myrica gale*: Photoproducts and effect of terpenes. *Journal of Agricultural and Food Chemistry*, 67(26), 7258–7265.

Kröber, M., Wibberg, D., Grosch, R., Eikmeyer, F., Verwaaijen, B., Chowdhury, S.P., Hartmann, A., Pühler, A., Schlüter, A. (2014). Effect of the strain *Bacillus amyloliquefaciens* FZB42 on the microbial community in the rhizosphere of lettuce under field conditions analyzed by whole metagenome sequencing. *Frontiers in Microbiology*, 5, 252.

Lankadurai, B.P., Nagato, E.G., Simpson, M.J. (2013). Environmental metabolomics: An emerging approach to study organism responses to environmental stressors. *Environmental Reviews*, 21(3), 180–205.

Larkin, R.P. (2016). Impacts of biocontrol products on Rhizoctonia disease of potato and soil microbial communities, and their persistence in soil. *Crop Protection*, 90, 96–105.

Lénárt, J., Bujna, E., Kovács, B., Békefi, E., Száraz, L., Dernovics, M. (2013). Metabolomic approach assisted high resolution LC–ESI-MS based identification of a xenobiotic derivative of fenhexamid produced by *Lactobacillus casei*. *Journal of Agricultural and Food Chemistry*, 61(37), 8969–8975.

Li, L., Zhou, Z., Jin, W., Wan, Y., Lu, W. (2015). A transcriptomic analysis for identifying the unintended effects of introducing a heterologous glyphosate-tolerant EPSP synthase into *Escherichia coli*. *Molecular BioSystems, The Royal Society of Chemistry*, 11(3), 852–858.

Li, L., Ma, J., Ibekwe, A.M., Wang, Q., Yang, C.-H. (2016). Cucumber rhizosphere microbial community response to biocontrol agent *Bacillus subtilis* B068150. *Agriculture*, 6(1), 2.

Li, M., Song, J., Ma, Q., Kong, D., Zhou, Y., Jiang, X., Parales, R., Ruan, Z., Zhang, Q. (2020). Insight into the characteristics and new mechanism of nicosulfuron biodegradation by a Pseudomonas sp. LAM1902. *Journal of Agricultural and Food Chemistry*, 68(3), 826–837.

Lin, C.Y., Viant, M.R., Tjeerdema, R.S. (2006). Metabolomics: Methodologies and applications in the environmental sciences. *Journal of Pesticide Science*, 31(3), 245–251.

Longnecker, K., Futrelle, J., Coburn, E., Kido Soule, M.C., Kujawinski, E.B. (2015). Environmental metabolomics: Databases and tools for data analysis. *Marine Chemistry*, 177, 366–373.

Lü, Z., Min, H., Wu, S., Ruan, A. (2003). Phylogenetic and degradation characterization of *Burkholderia cepacia* WZ1 degrading herbicide quinclorac. *Journal of Environmental Science and Health, Part B*, 38(6), 771–782.

Lu, W., Li, L., Chen, M., Zhou, Z., Zhang, W., Ping, S., Yan, Y., Wang, J., Lin, M. (2013). Genome-wide transcriptional responses of *Escherichia coli* to glyphosate, a potent inhibitor of the shikimate pathway enzyme 5-enolpyruvylshikimate-3-phosphate synthase. *Molecular BioSystems*, 9(3), 522–530.

Mallet, C., Romdhane, S., Loiseau, C., Béguet, J., Martin-Laurent, F., Calvayrac, C., Barthelmebs, L. (2019). Impact of Leptospermone, a natural β-Triketone herbicide, on the fungal composition and diversity of two arable soils. *Frontiers in Microbiology*, 10, 1024.

Martin-Laurent, F., Kandeler, E., Petric, I., Djuric, S., Karpouzas, D.G. (2013). ECOFUN-MICROBIODIV: An FP7 European project for developing and evaluating innovative tools for assessing the impact of pesticides on soil functional microbial diversity – Towards new pesticide registration regulation? *Environmental Science and Pollution Research*, 20(2), 1203–1205.

Meng, D., Zhang, L., Meng, J., Tian, Q., Zhai, L., Hao, Z., Guan, Z., Cai, Y., Liao, X. (2019). Evaluation of the strain *Bacillus amyloliquefaciens* YP6 in phoxim degradation via transcriptomic data and product analysis. *Molecules*, 24(21):3997.

Mesnage, R., Oestreicher, N., Poirier, F., Nicolas, V., Boursier, C., Vélot, C. (2020). Transcriptome profiling of the fungus *Aspergillus nidulans* exposed to a commercial glyphosate-based herbicide under conditions of apparent herbicide tolerance. *Environmental Research*, 182, 109116.

Morandin, L.A., Winston, M.L., Franklin, M.T., Abbott, V.A. (2005). Lethal and sub-lethal effects of spinosad on bumble bees (*Bombus impatiens* Cresson). *Pest Management Science*, 61(7), 619–626.

Namouchi, A., Gómez-Muñoz, M., Frye, S.A., Moen, L.V., Rognes, T., Tønjum, T., Balasingham, S.V. (2016). The *Mycobacterium tuberculosis* transcriptional landscape under genotoxic stress. *BMC Genomics*, 17(1), 791.

Nde, C.W., Jang, H.-J., Toghrol, F., Bentley, W.E. (2008). Toxicogenomic response of *Pseudomonas aeruginosa* to ortho-phenylphenol. *BMC Genomics*, 9(1), 473.

OECD (2000a). Test No. 216: Soil microorganisms: Nitrogen transformation test. Guideline, OECD, 10.

OECD (2000b). Test No. 217: Soil microorganisms: Carbon transformation test. Guideline, OECD, 10.

OECD (2019). Report of the 9th Biopesticides Expert Group Seminar on test methods for microorganisms (ENV/JM/MONO(2019)8). Report, OECD, 37.

O'Malley, M.A. (2013). Metametabolomics. In *Encyclopedia of Systems Biology*, 1st edition, Dubitzky, W., Wolkenhauer, O., Yokota, H., Cho, K.-H. (eds). Springer-Verlag, New York.

Oracz, K., Voegele, A., Tarkowská, D., Jacquemoud, D., Turečková, V., Urbanová, T., Strnad, M., Sliwinska, E., Leubner-Metzger, G. (2012). Myrigalone A inhibits *Lepidium sativum* seed germination by interference with gibberellin metabolism and apoplastic superoxide production required for embryo extension growth and endosperm rupture. *Plant and Cell Physiology*, 53(1), 81–95.

Patil, C., Calvayrac, C., Zhou, Y., Romdhane, S., Salvia, M.-V., Cooper, J.-F., Dayan, F.E., Bertrand, C. (2016). Environmental metabolic footprinting: A novel application to study the impact of a natural and a synthetic β-triketone herbicide in soil. *Science of the Total Environment*, 566–567, 552–558.

Pavela, R. and Benelli, G. (2016). Essential oils as ecofriendly biopesticides? Challenges and constraints. *Trends in Plant Science*, 21(12), 1000–1007.

Popovici, J., Bertrand, C., Bagnarol, E., Fernandez, M.P., Comte, G. (2008). Chemical composition of essential oil and headspace-solid microextracts from fruits of *Myrica gale* L. and antifungal activity. *Natural Product Research*, 22(12), 1024–1032.

Popovici, J., Bertrand, C., Comte, G. (2014). Use of a Myrica Gale plant for producing a herbicide agent [Online]. Available at: https://patents.google.com/patent/US8734858B2/en.

Pubmed.gov (2021a). Search for the term "Metabolomics" [Online]. Available at: https://pubmed.ncbi.nlm.nih.gov/?term=(metabolomics) [Accessed 23 June 2021].

Pubmed.gov (2021b). Search for the terms "Metabolomics" and "Environmental" [Online]. Available at: https://pubmed.ncbi.nlm.nih.gov/?term=(metabolomics)%20AND%20(environmental) [Accessed 23 June 2021].

Robin, D.C. and Marchand, P.A. (2019). Evolution of the biocontrol active substances in the framework of the European Pesticide Regulation (EC) No. 1107/2009: Evolution of BCA within the framework of Regulation (EC) No. 1107/2009. *Pest Management Science*, 75(4), 950–958.

Rodríguez, A., Castrejón-Godínez, M.L., Salazar-Bustamante, E., Gama-Martínez, Y., Sánchez-Salinas, E., Mussali-Galante, P., Tovar-Sánchez, E., Ortiz-Hernández, M.L. (2020). Omics approaches to pesticide biodegradation. *Current Microbiology*, 77(4), 545–563.

Romdhane, S., Devers-Lamrani, M., Barthelmebs, L., Calvayrac, C., Bertrand, C., Cooper, J.-F., Dayan, F.E., Martin-Laurent, F. (2016). Ecotoxicological impact of the bioherbicide leptospermone on the microbial community of two arable soils. *Frontiers in Microbiology*, 7:775.

Romdhane, S., Devers-Lamrani, M., Beguet, J., Bertrand, C., Calvayrac, C., Salvia, M.-V., Jrad, A.B., Dayan, F.E., Spor, A., Barthelmebs, L. et al. (2019). Assessment of the ecotoxicological impact of natural and synthetic β-triketone herbicides on the diversity and activity of the soil bacterial community using omic approaches. *Science of the Total Environment*, 651, 241–249.

Salvia, M.-V., Ben Jrad, A., Raviglione, D., Zhou, Y., Bertrand, C. (2018). Environmental metabolic footprinting (EMF) vs. half-life: A new and integrative proxy for the discrimination between control and pesticides exposed sediments in order to further characterise pesticides' environmental impact. *Environmental Science and Pollution Research*, 25(30), 29841–29847.

Santos, V.S.V., Silva, C.E., Oliveira, C.M., de Morais, C.R., Limongi, J.E., Pereira, B.B. (2019). Evaluation of toxicity and environmental safety in use of spinosad to rationalize control strategies against *Aedes aegypti*. *Chemosphere*, 226, 166–172.

Savazzini, F., Longa, C.M.O., Pertot, I. (2009). Impact of the biocontrol agent *Trichoderma atroviride* SC1 on soil microbial communities of a vineyard in northern Italy. *Soil Biology and Biochemistry*, 41(7), 1457–1465.

Schreiter, S., Babin, D., Smalla, K., Grosch, R. (2018). Rhizosphere competence and biocontrol effect of *Pseudomonas* sp. RU47 independent from plant species and soil type at the field scale. *Frontiers in Microbiology*, 9, 97.

Seo, J.-S., Keum, Y.-S., Li, Q.X. (2013). Metabolomic and proteomic insights into carbaryl catabolism by *Burkholderia* sp. C3 and degradation of ten N-methylcarbamates. *Biodegradation*, 24(6), 795–811.

Svoboda, K.P., Inglis, A., Hampson, J., Galambosi, B., Asakawa, Y. (1998). Biomass production, essential oil yield and composition of *Myrica gale* L. harvested from wild populations in Scotland and Finland. *Flavour and Fragrance Journal*, 13(6), 367–372.

Szewczyk, R., Soboń, A., Słaba, M., Długoński, J. (2015). Mechanism study of alachlor biodegradation by *Paecilomyces marquandii* with proteomic and metabolomic methods. *Journal of Hazardous Materials*, 291, 52–64.

Tamm, L., Schaerer, H.-J., Levert, A., Andreu, V., Bertrand, C. (2017). Agent antifongique, procédé et composition (FR1600099), 28 [Online]. Available at: https://bases-brevets.inpi.fr/fr/document/FR3046907.html [Accessed 10 June 2021].

Thiour-Mauprivez, C., Martin-Laurent, F., Calvayrac, C., Barthelmebs, L. (2019). Effects of herbicide on non-target microorganisms: Towards a new class of biomarkers? *Science of the Total Environment*, 684, 314–325.

Tiwari, B., Verma, E., Chakraborty, S., Srivastava, A.K., Mishra, A.K. (2018). Tolerance strategies in cyanobacterium *Fischerella* sp. under pesticide stress and possible role of a carbohydrate-binding protein in the metabolism of methyl parathion (MP). *International Biodeterioration & Biodegradation*, 127, 217–226.

Tomé, H.V.V., Barbosa, W.F., Martins, G.F., Guedes, R.N.C. (2015). Spinosad in the native stingless bee Melipona quadrifasciata: Regrettable non-target toxicity of a bioinsecticide. *Chemosphere*, 124, 103–109.

Tweeddale, H., Notley-McRobb, L., Ferenci, T. (1998). Effect of slow growth on metabolism of *Escherichia coli*, as revealed by global metabolite pool ("Metabolome") analysis. *Journal of Bacteriology*, 180(19), 5109–5116.

Vailati-Riboni, M., Palombo, V., Loor, J.J. (2017). What are omics sciences? In *Periparturient Diseases of Dairy Cows*, 1st edition, Ametaj, B.N. (ed.). Springer, Cham.

Viant, M.R. and Sommer, U. (2013). Mass spectrometry based environmental metabolomics: A primer and review. *Metabolomics*, 9(S1), 144–158.

Villaverde, J.J., Sevilla-Morán, B., Sandín-España, P., López-Goti, C., Alonso-Prados, J.L. (2014). Biopesticides in the framework of the European Pesticide Regulation (EC) No. 1107/2009: Biopesticides within the European Pesticide Regulation (EC) No. 1107/2009, *Pest Management Science*, 70(1), 2–5.

Wolfand, J.M., LeFevre, G.H., Luthy, R.G. (2016). Metabolization and degradation kinetics of the urban-use pesticide fipronil by white rot fungus *Trametes versicolor*. *Environmental Science: Processes & Impacts*, 18(10), 1256–1265.

Wu, Y. and Li, L. (2016). Sample normalization methods in quantitative metabolomics. *Journal of Chromatography A*, 1430, 80–95.

Wu, S.C., Cao, Z.H., Li, Z.G., Cheung, K.C., Wong, M.H. (2005). Effects of biofertilizer containing N-fixer, P and K solubilizers and AM fungi on maize growth: A greenhouse trial. *Geoderma*, 125(1–2), 155–166.

4

Plant Secondary Metabolites Mode of Action in the Control of Root-Knot Nematodes

Nikoletta NTALLI[1] and Pierluigi CABONI[2]

[1] *Department of Agriculture Crop Production and Rural Environment, University of Thessaly, Greece*

[2] *Department of Life and Environmental Sciences, University of Cagliari, Italy*

Meloidogyne spp. constitute major agricultural pests that are scarcely controlled due to the shortage and misfunctioning of commercial nematicides, which additionally exhibit detrimental effects for the environment and humans. In fact, the use of "old-generation" synthetic nematode control agents is frequently banned for reasons of environmental safety. In the search for effective and ecofriendly alternatives to their synthetic ancestors, the naturally originating nematicidals have been the focus of research in recent years. Botanical extracts are proved of nematicidal activity, and constituent natural compounds are chemically characterized with robust analytical methods as active ingredients. The efficacy is usually addressed in small-scale in vitro and pot bioassays, where the frequently cited assessment parameters are the second-stage (J2) juveniles' paralysis, the egg hatch inhibition and the parasite's biological cycle arrest in host roots. Yet, the biochemical basis of the activity, such as the molecular targets inside the nematode where the compound interacts, has only seldom been reported for the natural nematicidals. Herein, we summarize the literature published in the last decade, on the functional and anatomical changes that are induced on different root-knot nematodes' growth stages exposed

Biocontrol of Plant Disease,
coordinated by Claire PRIGENT-COMBARET and Bernard DUMAS. © ISTE Ltd 2022.

to plant secondary metabolites, along with any intrinsic biochemical interactions supporting the effect.

4.1. Introduction

The need for alternative nematode control measures has grown increasingly in recent years due to environmental, food safety and animal welfare issues related to the synthetic nematicides. A non-chemical nematode control method, mostly practiced to combat root-knot nematodes in glasshouses, is soil sterilization, where the soil is steamed every five to six production cycles. But this technique is expensive, environmentally unfriendly and reduces the resistance and resilience of soil against pathogens and pests (De Long et al. 2021). On the other hand, the natural substances originating from botanical matrices and involved in the complex chemical-mediated interactions between a plant and other organisms in its environment can represent significant candidates for developing innovative nematode control tools that can be safely incorporated into integrated pest management programs (Ntalli and Caboni 2012). In this frame, some botanicals have reached the pipeline of authorization in the EU, among which is a formulate based on garlic, *Allium sativum* named NEMguard®. In an in vitro evaluation of the lethal concentration, NEMguard® exhibited a strong nematicidal activity. The LC_{50} value was calculated at 0.8 mg/mL, after 96 h of second-stage juvenile (J2) immersion in test wells, and the respective LC_{90} was calculated at 1.5 mg/mL. The product showed a protective effect against *M. incognita* as a single application in small pots, while a second application helped augment the reduction of root galling significantly. Nonetheless, when NEMguard® was applied in the greenhouse, singularly or monthly, no control was observed on *M. incognita*, probably due to the silt content of the loamy sandy soil having an effect on the contained polysulfides of the formulate (Eder et al. 2021).

Here, we present the international bibliography cited in Scopus® in the last 10 years considering the botanical nematicidals used to combat root-knot nematodes. We separate the findings according to the efficacy evaluation scale, that is, in vitro and in vivo, as well as the biochemical mode of action, if reported.

4.2. Recent research on the use of plant secondary metabolites to control *Meloidogyne* spp.

Research on using plants as control agents for the plant-parasitic nematodes has received substantial attention in recent years. Small-scale in vitro experiments are used as screening bioassays to investigate the nematicidal activity of botanical extracts or plant secondary metabolites. The efficacy is assessed on the basis of juvenile's mortality or egg hatchability inhibition after immersion in water test solutions. The larger-scale in vivo, glasshouse and field experiments usually involve powdered meal (plant material) or extracts, incorporated in the soil or used for seed treatment to interfere with the parasites' biological cycle in host roots. Further research indicating constituent components mode of action is warranted (Makhubu et al. 2021).

4.2.1. *Small-scale experiments reporting in vitro efficacy*

Small-scale experiments are employed either involving extracts or constituent components tested individually or in artificial mixes. Herewith are the most recently reported nematicidal plant extracts combating root-knot nematodes followed by the lately discovered nematicidal botanical individual compounds. Kesba et al. (2021) tested solidago and periwinkle (*Vinca minor*) aqueous and ethanolic extracts on the root-knot nematode, *Meloidogyne incognita* J2, achieving greater mortality with the aqueous extracts. In addition, the inhibition of egg hatching exhibited by the periwinkle extracts was higher than that of solidago (Kesba et al. 2021). In an egg hatch inhibition activity study of *M. incognita*, the crude extracts of *Cucurbita maxima*, *Prunus africana*, *Pelargonium sidoides*, *Croton sylvaticus*, *Solanum aculeastrum*, *Vernonia colorata*, *Searsia lancea* and *Merwilla plumbea* were employed and all eight plant extracts showed promising nematicidal activity at 0.8 mg/mL (Sithole et al. 2021). Marigold (*Tagetes* spp.) and cabbage (*Brassica oleracea*) are two widely studied botanicals having shown their potential to control root-knot nematodes, in the form of organic amendments and botanical extracts. When tested for the mortality of J2 and egg hatch inhibition of *M. javanica*, the activity was higher and steadier for the botanical extracts rather than for the organic amendments (Das et al. 2021). Different plant parts' extracts of *Acacia modesta*, *Segeratia thea* and *Celtis caucasic* paralyzed *M. incognita* J2 and gradually decreased egg hatching (Humaira et al. 2020). *Coccinia grandis*, *Commelina benghalensis*, *Leucas cephalotes*, *Phyllanthus amarus* and

Trianthema portulacastrum extracts provoked paralysis and repressed *M. incognita* egg hatch (Khan et al. 2019). *Artemisia pedemontana* essential oil was proved nematicidal against *M. javanica* (Sainz et al. 2019). *Camellia oleifera* cake and stems from two tree peony species, *Paeonia rockii* and *Paeonia suffruticosa*, were antagonistic to *M. incognita* in in vitro experiments. In particular, extracts from *C. oleifera* cake and *P. rockii* stems blocked hatch and were nematotoxic to J2, and *P. rockii* extracts were more effective (Wen et al. 2019). In another study, the essential oil and hydrosol of cumin, *Cuminum cyminum* seeds, were evaluated against *M. incognita* and *M. javanica* in terms of J2s paralysis, as well as egg differentiation and the essential oil was found to be more active (Pardavella et al. 2021). Hydrolates from *Lavandula intermedia*, *Lavandula luisieri*, *Thymus vulgaris* and *Thymus zygis* showed strong in vitro nematicidal effects against *M. javanica* in terms of J2 mortality and suppression of egg hatching (Andrés et al. 2018). Hydrosols of *Origanum vulgare*, *Mentha piperita* and *Melissa officinalis*, as well as water extracts of *Origanum vulgare*, *Thymus vulgaris*, *Thymus citriodorus*, *Rosmarinus officinalis* and *Ocimum basilicum* were found to be active against *M. javanica* and *M. incognita* (Ntalli et al. 2020a). In the same frame, the essential oil and hydrosol of *Satureja hellenica* paralyzed the J2 and inhibited the egg hatch of *M. incognita* and *M. javanica* (Pardavella et al. 2020). When the nematicidal efficacy of the aqueous and methanolic extracts of *Peganum harmala*, *Raphanus raphanistrum*, *Taxus baccata*, *Sinapis arvensis* and *Ricinus communis* was evaluated on *M. incognita* J2 paralysis, the methanolic extracts were found to be significantly more active. The best effective methanolic extracts were those of *T. baccata*, *P. harmala* and *S. arvensis*, followed by the aqueous extract of *T. baccata* (Zaidat et al. 2020). Aqueous extracts of crambe, *Crambe abyssinica* seed cake, prepared at different temperatures reduced egg hatching and increased juvenile mortality of *M. javanica* (Tarini et al. 2020). The aqueous extracts of silver nightshade, *Solanum elaeagnifolium*, exhibited strong impacts on *M. incognita* galls, egg-masses, the number of females and the number of juveniles (Balah and AbdelRazek 2020). In another study, the essential oils of *Ocimum sanctum*, *Cymbopogon schoenanthus* and *Cinnamomum zeylanicum* showed a great potential against *M. incognita*. Among the constituent compounds identified were cinnamyl acetate, methyl eugenol, cinnamyl alcohol, acetyl eugenol, isoeugenol, eugenol and benzyl benzoate. Additionally, a synergistic nematicidal activity was evident when phenylpropanoids were combined with carvone (Eloh et al. 2020). The constituent hallerone found in *Digitalis obscura* showed strong nematicidal effects on *Meloidogyne javanica* juveniles (Quílez del Moral et al. 2021). D'Addabbo et al. (2020) proved that saponins' rich *Medicago heyniana*, *Medicago hybrida*, *Medicago lupulina*, *Medicago murex* and *Medicago truncatula* controlled *M. incognita* both in terms of J2 paralysis and egg hatch inhibition. Hexanoic acid and acetic acid, two active ingredients of the nematicidal *Melia azedarach*, when used singularly provoked paralysis activity on second-stage juveniles and significantly inhibited *M. javanica* undifferentiated egg hatch and J2 release from free eggs immersed in test solutions (Ntalli et al. 2020b). In another study, *Tagetes patula* extract, rich in alkaloids, terpenoids, steroids, glycosides, tannins and saponins, showed inhibitory activity towards egg hatch of *M. incognita* (Aiyelaagbe et al. 2020). *Croton ehrenbergii* yielded phenylalanine, l-quebrachitol, trans-4-hydroxy-*N*-methylproline and the flavonoids: kaempferol, tiliroside, nicotiflorine and rutin with potential nematicidal properties against *M. incognita* (Rios et al. 2019). Thiophenes are sulfur atom molecules of botanical origin, naturally existing in wormwood, *Artemisia absinthium*, and are of significant nematicidal properties (Liu et al. 2019a). These compounds can also be isolated from *Echinops grijsii*, and an increased number of acetylenes and chlorine atoms in their structure promotes their activity (Liu et al. 2019b). Camphor and 2,3,4,4-tetramethyl-5-methylidenecyclopent-2-en-1-one, being the major components of the hydrolate of *Lavandula luisieri*, displayed significant nematicidal properties without harming non-target organisms (Pino-Otín et al. 2019). Phenanthrene derivatives (parviphenanthrines A–F), stilbene derivatives (parvistilbines A–B), esters (parviesters A–C) and sesquiterpenoids (parvidiol A) naturally occurring in *Stemona parviflora* roots controlled *M. incognita* (Huang et al. 2019). Aristolochic acid I, aristololactam I and aristololactam W from the methanol extract of *Aristolochia mollissima* paralyzed *M. javanica* J2 (Bu et al. 2018). *Eugenia winzerlingii* hexane and ethyl acetate extracts, rich in decanoic, undecanoic, and dodecanoic acids, were found nematicidal against *M. incognita* and *M. javanica* (Cruz-Estrada et al. 2019).

4.2.2. *Large-scale experiments reporting in vivo efficacy*

Plant residues of *Raphanus sativus* (radish) and *Oryza sativa* (rice) soil amendments demonstrated up to 85% reduction in host root damage and root-knot nematode population densities, while leaf extracts of *Allium tuberosum* (garlic) and leaf powder of *Laurelia sempervirens* (Peruvian nutmeg) established the highest efficacy (Peiris 2021). *Meloidogyne* sp. population, root-knot index and soil egg hatching were reduced in pot trials employing plant parts of fermented *Lantana camara* L. as soil amendments to fight against root-knot nematodes (Malahlela et al. 2021). Nemafric-BL is a cucurbitacin-containing phytonematicide, prepared from fruits dried at 52°C in air-forced ovens or from fermented crude extracts of fresh fruit from wild watermelon (*Cucumis africanus*) (Tseke and Mashela 2017). Nemafric-BL applied once weekly for eight weeks at 4.03% managed *M. javanica* population densities in *Sutherlandia frutescens* (Masenya et al. 2020). Different plant part extracts from *Echinacea angustifolia*, *Echinacea pallida* and *Echinacea purpurea* were found nematidical on juveniles and eggs of *M. incognita*. Polar compounds such as echinacoside and chicoric acid were the main components of the extracts from aerial parts, whereas apolar components such as alkamides prevailed in root extracts (Laquale et al. 2020). The lemon thyme powder, macerate, water extract and essential oil were proved to control *M. incognita* and *M. javanica* in containerized tomato plants (Ntalli et al. 2020c). In another study, cinnamon bark extracts and chitosan were active against *M. incognita* and *Pratylenchus coffeae* under pot and field conditions (Nguyen et al. 2020). According to Jaafar et al. (2020), *Raphanus sativus* has a high ability to control nematodes present in the soil, such as *Meloidogyne hapla*, and is thus preferable to include in agricultural cycles in areas endemic to this nematode. In another study, the aqueous extracts of *Pistacia lentiscus* controlled *Meloidogyne javanica* under greenhouse conditions (Hajji-Hedfi et al. 2019). *Waltheria indica,* in the form of a wettable powder-type formulation of the ethyl acetate extract, exhibited a strong nematicidal activity in a pot experiment against *Meloidogyne incognita* in the form of a wettable powder-type formulation of the ethyl acetate extract. Three 4-quinolone alkaloids, namely 5′-methoxywaltherione A, waltherione A and waltherione C, were isolated and characterized as nematicidal (Jang et al. 2019). In another study, *Amygdalus scoparia* leaf extract at the rate of 2% reduced the number of galls, egg masses and eggs per root system, as well as the number of J2s per pot and the reproduction factor of nematode in a pot experiment (Rafiee et al. 2019). Extracts of *Curcuma rhizome, Ginger rhizome* and pumpkin seeds caused reduction in *M. javanica* penetration and development in eggplant root tissues (Al-Sandooq et al. 2019). Treatment of *M. incognita* affected tomato plants with chopped leaves of turmeric (*Curcuma longa*), marwatulsi (*Origanum majorana*), mentha (*Mentha arvensis*), aonla (*Phyllanthus emblica*) and jatropha (*Jatropha curcas*), and reduced reproduction factors such as the number of galls/plant, number of egg masses/plant and final nematode population in soil (Neeraj et al. 2019). Powdered and aqueous extract of citrus, rich in citral and *β*-caryophyllene, tannins, saponins, flavonoids, glycosides, sterols and alkaloids, controlled root-knot nematode infecting cowpea (Izuogu et al. 2019). Cabbage leaves and faba bean pod plant waste-hydro-ethanolic extracts exhibited hatching inhibition and repellent activity, as well as population development and *M. incognita* infection inhibition on the tomato root system (Roshan-Bakhsh et al. 2019). Similarly, aqueous extracts of the agro-industry waste coffee silverskin and brewers' spent grain effectively controlled *M. incognita* in tomato plants (Thligene et al. 2019). Likewise, olive pomace and green walnut husk soil amendments, rich in 5-hydroxy-1,4-naphthoquinone, controlled *M. incognita* in tomato (Kavdir et al. 2019). Cucurbitacin-containing phytonematicides were better nematicidal with respect to their purified active ingredients cucurbitacins (Dube et al. 2019). Neem seed kernel extract, neem, aak and castor leaves not only had a suppressive effect on the root-knot nematode, but also enhanced the plant growth parameters, supplying nutritional elements to cucumber plants (Patil et al. 2018). *Cleome viscosa* shoot powder, rich in phenols, alkaloids, tannins and terpenoids, encouraged better plant growth and reduced nematode populations in okra (Claudius-Cole et al. 2018). Dry leaf and root biomass of *Taraxacum officinale* applied to *M. incognita* infested soil strongly suppressed nematode multiplication and gall formation. The chicoric acid and 3-*O*- and 3,5-di-*O*-caffeoylquinic acids (Laquale et al. 2018a) were identified as the major components. In pot experiments, hydrolates of *Lavandula intermedia* var. super, *Lavandula luisieri*, *Thymus vulgaris* and *Thymus zygis* tested on tomato plants significantly affected the infection frequency and reproduction rate of the *M. javanica* population, thus rendering them an exploitable source of potential waste protection products on root-knot nematodes (Andrés et al. 2018). Different crude protein extracts precipitating from *Moringa oleifera* seeds, controlled *M. incognita*, on banana cv. Grande-Naine plants (El-Ansary and Al-Saman 2018). Margosa, marigold, datura, hemp and tobacco extract treated and nematode inoculated plants showed improvement on growth over untreated inoculated plants and reduced the development of disease. Margosa and marigold leaf extracts were found to be the most effective in controlling the root-knot development (Shah et al. 2018). Neem cake, sesame oil and quillay extracts significantly reduced nematode multiplication and gall formation on tomato roots. Neem cake and sesame oil were significantly more suppressive than the quillay extract formulation, while the algae formulation provided the lowest effect on nematode infestation (Laquale et al. 2018b).

4.2.3. *Mechanism of action*

The research in the biochemical mode of action of plant secondary metabolites against plant pathogenic nematodes has yet to become extensive.

The most recent findings on the topic are reported hereafter. In a scanning electron microscope study, biochar-coated urea treatment for 12 h led to the cuticle surface of J2 losing its normal aspect, showing a crimped appearance with distorted anus (Abdelnabby et al. 2018). In another study, acetylcholinesterase activity was suppressed by more than 75% in *Helicotylenchus dihystera* treated with the extract of *Punica granatum*, *Thymus vulgaris* and *Artemisia absinthium* (Korayem et al. 1993). *Moringa oleifera* seed exudates yielded defense proteins such as β-1,3-glucanases, chitinases, proteases, serine and cysteine, protease inhibitors of significant anthelmintic activity. The exuded proteins presented ovicidal activity and caused 100% mortality of *M. incognita* J2s. Scanning electron microscopy provoked invaginations, cracks, scratched surface and loss of internal content on *M. incognita* eggs. Other than the anthelmintic proteins in *M. oleifera* seed exudate, it is possibly involved in plant defense mechanisms during seed germination (Sousa et al. 2020). In another study, *Metarhizium guizhouense* spores suspended in crude protease rapidly germinated in 24 h and then mycelia managed to invade immature eggs of *M. incognita*, while the eggs containing juveniles and the hatched juveniles were resistant. Scanning electron microscopy observations showed widespread injury to eggshell and cuticle when treated with *Metarhizium* spores and protease, either individually or in combination (Thongkaewyuan and Chairin 2018). *Ficus carica* extracts exhibited strong paralysis activity against *M. javanica* J2, while microscopy analysis revealed severe morphological alterations in the nematode tissues at the J2 stage (Alves et al. 2020). Aqueous extracts of *Allium sativum*, *Urtica dioica*, *Sophora mollis*, *Ephedra intermedia* and *Tanacetum baltistanicum* were nematicidal against *M. incognita*, demonstrating that some polar oxygenated secondary metabolites with lipophilic properties may be responsible for the damage of the cytoplasmic membrane of the nematode cells by interfering with the enzyme protein structure through their functional groups (Ismail et al. 2020). Wang et al. (2018) studied the behaviors of *Heterodera glycines* and *M. incognita* in the presence of root signals from marigold (*Tagetes patula*), soybean (*Glycine max*) and pepper (*Capsicum annuum*). Root tips from each plant species attracted *M. incognita*, but *H. glycines* was attracted only to soybean. In contrast, root exudates prepared from marigold, pepper or soybean seedlings were attractive to *H. glycines*, but were repellent to *M. incognita*. Root extracts had the same effect as exudates (Wang et al. 2018). In another study, they showed that, when grown with Welsh onion as a companion plant, cucumber roots had 77.0% fewer root knots and egg masses than the control cucumber roots. Treatment with 1.2 mM 4-hydroxy-benzeneethanol decreased egg hatchability along with the activity of the J2 (Li et al. 2018).

4.3. Conclusion

For its major part, the bibliography on botanicals controlling root-knot nematodes represents small-scale efficacy data without further testing under field conditions. The mode of action is yet to be delineated for many botanical secondary metabolites exhibiting nematicidal properties, as well as extracts composing bioactive complex clusters of compounds. In fact, the mode of action of composite plant extracts remains vague for its major part, although mandatory for registration purposes. Along with the efficacy data, the toxicity and ecotoxicity are also to be outlined because being of natural origin does not render the botanical bioactive substances safe. The EU commission's commitment to tackling climate and environmental-related challenges is increasingly defining new tasks regarding the use of pesticides. In 2019, the Commission adopted the Communication on "The European Green Deal". Botanicals may represent significant candidates for ecofriendly alternatives to the synthetic pesticides and they could be of use for the farmer either if developed into low-risk plant protection products (article 22 and 47 of Reg. (EC) No. 1107/2009) or as basic substances (article 23 of Reg. (EC) No. 1107/2009) (for these articles, see European Parliament (2009)).

4.4. References

Abdelnabby, H., Hu, Z., Wang, H., Zhang, X. (2018). Furfural–biochar-based formulations show synergistic and potentiating effects against *Meloidogyne incognita* in tomato. *J. Pest Sci.*, 91, 203–218.

Aiyelaagbe, O., Olaide, O., Claudius-Cole, A., Aiyelaagbe, I., Ojo, O. (2020). Egg hatch inhibition of root-knot nematodes by *Tagetes patula* extracts. *Acta Hortic.*, 1269, 129–134.

Al-Sandooq, D.L., Mutleb, W.K., Abbas, M.K. (2019). Inhibition activity of curcuma, ginger and pumpkin seeds extracts against root knot nematodes on eggplant. *Plant Arch.*, 19(2), 3946–3950.

Alves, J.R., Nunes de Assis, J., Pádua, C.C.A., Balbino, H.M., Lima, L.L., de Souza Gouveia, A., Vital, C.E., Buonicontro, D.S., de Freitas, L.G., Leite, J.P.V. et al. (2020). *J. Plant Prot. Res.*, 60(2), 192–206.

Andrés, M.F., González-Coloma, A., Muñoz, R., De la Peña, F., Julio, L.F., Burillo, J. (2018). Nematicidal potential of hydrolates from the semi-industrial vapor-pressure extraction of Spanish aromatic plants. *Environ. Sci. Pollut. Res.*, 25(30), 29834–29840.

Balah, M.A. and AbdelRazek, G.M. (2020). Pesticidal activity of *Solanum elaeagnifolium* Cav. leaves against nematodes and perennial weeds. *Acta Ecol. Sin.*, 40(5), 373–379.

Bu, M., Yu, S., Dong, C. (2018). Chemical constituents from fruits of *Aristolochia mollissima* and their nematicidal activity against root-knot nematode. *Zhongguo Zhong Yao Za Zhi*, 43(16), 3307–3314.

Claudius-Cole, A.O., Okayi, H.C., Wodoyin, R.O.A. (2018). Potential of *Cleome viscosa* L. for the management of root-knot nematodes (*Meloidogyne incognita*) in okra (*Abelmoschus esculentus* (L.) moench). *Trop. Agric.*, 95(4), 313–325.

Cruz-Estrada, A., Ruiz-Sánchez, E., Cristóbal-Alejo, J., González-Coloma, A., Andrés, M.F., Gamboa-Angulo, M. (2019). Medium-chain fatty acids from *Eugenia winzerlingii* leaves causing insect settling deterrent, nematicidal, and phytotoxic effects. *Molecules*, 24(9), 1724.

D'Addabbo, T., Argentieri, M.P., Żuchowski, J., Biazzi, E., Tava, A., Oleszek, W., Avato, P. (2020). Activity of saponins from *Medicago* species against phytoparasitic nematodes. *Plants*, 9(4), 443.

Das, S., Wadud, A., Khokon, M.A.R. (2021). Evaluation of the effect of different concentrations of organic amendments and botanical extracts on the mortality and hatching of *Meloidogyne javanica*. *Saudi J. Biol. Sci.*, 28(7), 3759–3767.

De Long, J.R., Streminska, M.A., Persijn, A., Huisman, H.M.I., Van Der Salm, C. (2021). Biological control of *Meloidogyne* spp. in glasshouse-grown chrysanthemum. *J. Nem.*, 52, e2020-125.

Dube, Z.P., Mashela, P.W., de Waele, D. (2019). Sensitivity of *Meloidogyne incognita* second-stage juvenile hatch, motility and viability to pure cucurbitacins and cucurbitacin-containing phytonematicides. *S. Afr. J. Plant Soil*, 36(1), 29–32.

Eder, R., Consoli, E., Krauss, J., Dahlin, P. (2021). Polysulfides applied as formulated garlic extract to protect tomato plants against the root-knot nematode *Meloidogyne incognita*. *Plants*, 10(2), 394.

El-Ansary, M.S.M. and Al-Saman, M.A. (2018). Appraisal of *Moringa oleifera* crude proteins for the control of root-knot nematode, *Meloidogyne incognita* in banana. *Rend. Fis. Acc. Lincei*, 29(3), 631–637.

Eloh, K., Kpegba, K., Sasanelli, N., Koumaglo, H.K., Caboni, P. (2020). Nematicidal activity of some essential plant oils from tropical West Africa. *Int. J. Pest Manag.*, 66(2), 131–141.

European Parliament (2009). Regulation (EC) No. 1107/2009 of the European Parliament and of the council of 21 October 2009 concerning the placing of plant protection products on the market and repealing Council Directives 79/117/EEC and 91/414/EEC.

Hajji-Hedfi, L., Larayedh, A., Hammas, N.-C., Regaieg, H., Horrigue-Raouani, N. (2019). Biological activities and chemical composition of *Pistacia lentiscus* in controlling *Fusarium* wilt and root-knot nematode disease complex on tomato. *Eur. J. Plant Pathol.*, 155, 281–291.

Huang, S., Kong, F., Chen, G., Cai, X., Zhou, L., Ma, Q., Zhao, Y. (2019). A phytochemical investigation of stemona *parviflora* roots reveals several compounds with nematocidal activity. *Phytochemistry*, 159, 208–215.

Humaira, Shad, A.A., Muhammad, D., Shah, H.U. (2020). Evaluation of some medicinal plant extracts for their nematicidal properties against root-knot nematode, *Meloidogyne incognita. Appl. Ecol. Environ. Res.*, 18(2), 2475–2482.

Ismail, M., Fayyaz, S., Kowsar, A., Javed, S., Ali, I., Ali, S., Hussain, F., Ali, H. (2020). Evaluation of nematocidal effects of some medicinal plant extracts against root-knot nematodes (*Meloidogyne incognita*). *Ital. J. Agron.*, 15(1), 1475.

Izuogu, N.B., Azubuike, C., Hinmikaiye, A.S., Olarenwaju, A. (2019). Comparative evaluation of powdered and aqueous extract of citrus peels and carbofuran in the management of root-knot nematode on infected cowpea field. *Agric. Conspec. Sci.*, 84(2), 173–178.

Jaafar, N.A., Ahmed, A.S., Al-Sandooq, D.L. (2020). Detection of active compounds in radish *Raphanus sativus* l and their various biological effects. *Plant Arch.*, 20, 1647–1650.

Jang, J.Y., Dang, Q.L., Choi, G.J., Park, H.W., Kim, J-C. (2019). Control of root-knot nematodes using *Waltheria indica* producing 4-quinolone alkaloids. *Pest Manag. Sci.*, 75(8), 2264–2270.

Kavdir, Y., Gozel, U., Sahiner, N. (2019). Nematicidal effects of olive pomace and green walnut husk on root-knot nematode *Meloidogyne incognita* on tomato. *Allelopathy J.*, 46(1), 3–16.

Kesba, H., Abdel-Rahman, A., Sayed, S., Al-Sayed, A.-S. (2021). Screening the nematicidal potential of indigenous medicinal plant extracts against *Meloidogyne incognita* under lab and greenhouse conditions. *Egypt. J. Biol. Pest Control*, 31(1), 81.

Khan, F., Asif, M., Khan, A., Tariq, M., Ansari, T., Shariq, M., Siddiq, M.A. (2019). Evaluation of the nematicidal potential of some botanicals against root-knot nematode, *Meloidogyne incognita* infected carrot: *In vitro* and greenhouse study. *Curr. Plant Biol.*, 20, 100115.

Korayem, A.M., Hasabo, S.A., Ameen, H.H. (1993). Effects and mode of action of some plant extracts on certain plant parasitic nematodes. *Anz. Schadlingskde., Pflanzenschutz, Umweltschutz*, 66, 32–36.

Laquale, S., Avato, P., Argentieri, M.P., Candido, V., Perniola, M., D'Addabbo, T. (2018a). Nematicidal potential of taraxacum officinale. *Environ. Sci. Pollut. Res.*, 25(30), 30056–30065.

Laquale, S., Candido, V., D'Addabbo, T. (2018b). Side effects of biostimulants against root-knot nematodes on tomato. *Acta Hortic.*, 1207, 223–228.

Laquale, S., Avato, P., Argentieri, M.P., Candido, V., Perniola, M., D'Addabbo. T. (2020). Nematicidal activity of *Echinacea* species on the root-knot nematode *Meloidogyne incognita*. *J. Pest Sci.*, 93, 1397–1410.

Li, T., Wang, H., Xia, X., Cao, S., Yao, J., Zhang, L. (2018). Inhibitory effects of components from root exudates of Welsh onion against root knot nematodes. *PLoS One*, 13(7), e0201471.

Liu, T., Wu, H., Wu, H., Zhang, J. (2019a). Wormwood (*Artemisia absinthium* L.) as a promising nematicidal and antifungal agent: Chemical composition, comparison of extraction techniques and bioassay-guided isolation. *Ind. Crops Prod.*, 133, 295–303.

Liu, T., Wu, H., Jiang, H., Zhang, L., Zhang, Y., Mao, L. (2019b). Thiophenes from *Echinops grijsii* as a preliminary approach to control disease complex of root-knot nematodes and soil-borne fungi: Isolation, activities, and structure-nonphototoxic activity relationship analysis. *J. Agr. Food Chem.*, 67(22), 6160–6168.

Makhubu, F.N., Khosa, M.C., McGaw, L.J. (2021). South African plants with nematicidal activity against root-knot nematodes: A review. *S. Afr. J. Bot.*, 139, 183–191.

Malahlela, M., Thibane, V.S., Mudau, F.N. (2021). Nematocidal activity of fermented extracts from *Lantana camara* plant parts against *Meloidogyne javanica* on tomato. *Int. J. Veg. Sci.*, 27(1), 20–28.

Masenya, T.A., Pofu, K.M., Mashela, P.W. (2020). Responses of cancer bush (*Sutherlandia frutescens*) and *Meloidogyne javanica* to increasing concentration of Nemafric-BL phytonematicide. *Res. Crops*, 21(3), 615–620.

Neeraj, N., Kumar, A., Kumar, V., Kumar, M., Sindhu, A.S., Goel, S.R., Singh, G. (2019). Bio-nematicidal effect of botanicals against root-knot nematode (*Meloidogyne incognita*) in tomato (*Solanum lycopersicum*). *Indian J. Agric. Sci.*, 89(12), 2096–2101.

Nguyen, D.M.C., Luong, T.H., Nguyen, T.-K., Jung W.-J. (2020). Nematicidal activity of cinnamon bark extracts and chitosan against *Meloidogyne incognita* and *Pratylenchus coffeae*. *Nematol.*, 41(6), 1–12.

Ntalli, N. and Caboni, P. (2012). Botanical nematicides in the Mediterranean basin. *Phytochem. Rev.*, 11(4), 351–359.

Ntalli, N., Bratidou Parlapani, A., Tzani, K., Samara, M., Boutsis, G., Dimou, M., Menkissoglu-Spiroudi, U., Monokrousos. N. (2020a). *Thymus citriodorus* (Schreb) botanical products as ecofriendly nematicides with bio-fertilizing properties. *Plants*, 9(2), 202.

Ntalli, N., Menkissoglu-Spiroudi, U., Doitsinis, K., Kalomoiris, M., Papadakis, E-N., Boutsis, G., Dimou, M., Monokrousos. N. (2020b). Mode of action and ecotoxicity of hexanoic and acetic acids on *Meloidogyne javanica*. *J. Pest Sci.*, 93, 867–877.

Ntalli, N., Ozalexandridou, E.X., Kasiotis, K.M., Samara, M., Golfinopoulos, S.K. (2020c). Nematicidal activity and phytochemistry of Greek lamiaceae species. *Agronomy*, 10(8), 1119.

Pardavella, I., Nasiou, E., Daferera, D., Trigas, P., Giannakou, I. (2020). The use of essential oil and hydrosol extracted from satureja *hellenica* for the control of *Meloidogyne incognita* and *M. Javanica*. *Plants*, 9(7), 856.

Pardavella, I., Daferera, D., Tselios, T., Skiada, P., Giannakou, I. (2021). The use of essential oil and hydrosol extracted from *Cuminum cyminum* seeds for the control of *Meloidogyne incognita* and *Meloidogyne javanica*. *Plants*, 10(1), 46.

Patil, J., Kumar, A., Yadav, S., Goel, S.R., Bhatia, A.K. (2018). Bio-efficacy of phytotherapeutic substances against *Meloidogyne incognita* and *Fusarium oxysporum* F. sp. cucumerinum affecting cucumber in polyhouse under protected cultivation. *Indian J. Nematol.*, 48(2), 190–197.

Peiris, P.U.S. (2021). Use of botanicals in root-knot nematode control: A meta-analysis. *J. Plant Dis. Prot.*, 128, 913–922.

Pino-Otín, M.R., Val, J., Ballestero, D., Navarro, E., Sánchez, E., González-Coloma, A., Mainar, A.M. (2019). Ecotoxicity of a new biopesticide produced by *Lavandula luisieri* on non-target soil organisms from different trophic levels. *Sci. Total Env.*, 671, 83–93.

Quílez del Moral, J.F., Pérez, A., Navarro, M.J.S., Galisteo, A., Gonzalez-Coloma, A., Fe Andrés, M., Barrero. A.F. (2021). Selective extraction of bioactive phenylethanoids from *Digitalis obscura*. *Plants*, 10(5), 959.

Rafiee, F., Charehgani, H., Abdollahi, M. (2019). Evaluation of burdock and mountain almond leaf extracts against *Meloidogyne javanica* on tomato. *Arch. Phytopathol. Plant Path.*, 52, 11–12.

Rios, M.Y., León-Rivera, I., Ríos-Gomez, R., Córdova-Albores, L.C., Aguilar-Guadarrama, A.B. (2019). Phytotoxic and nematicide evaluation of *Croton ehrenbergii* (euphorbiaceae). *Pest Manag. Sci.*, 75(8), 2158–2165.

Roshan-Bakhsh, A., Pourjam, E., Ayyari, M., Pedram, M. (2019). Biocontrol properties of some agricultural waste extracts on three nematode species in *in vitro* and *in vivo* conditions. *Nematology*, 45, 1–10.

Sainz, P., Fe Andrés, M., Martínez-Díaz, R.A., Bailén, M., Navarro-Rocha, J., Díaz, C.E., González-Coloma, A. (2019). Chemical composition and biological activities of *Artemisia pedemontana* subsp. assoana essential oils and hydrolate. *Biomolecules*, 9(10), 558.

Shah, N.H., Dar, A.R., Qureshi, I.A., Akhter, A., Wani, M.R., Andleeb, L. (2018). Control of root-knot disease of brinjal (*Solanum melongena* L.) by the application of leaf extracts of certain medicinal plants. *Indian J. Agric. Res.*, 52(4), 444–447.

Sithole, N.T., Kulkarni, M.G., Finnie, J.F., Van Staden, J. (2021). Potential nematicidal properties of plant extracts against *Meloidogyne incognita*. *S. Afr. J. Bot.* 139, 409–417.

Sousa, A.J.S., Souza, P.F.N., Gifoni, J.M., Dias, L.P., Freitas, C.D.T., Oliveira, J.T.A., Sousa, D.O.B., Vasconcelos, I.M. (2020). Scanning electron microscopy reveals deleterious effects of *Moringa oleifera* seed exuded proteins on root-knot nematode *Meloidogyne incognita* eggs. *Int. J. Biol. Macromol.*, 154(1), 1237–1244.

Tarini, G., Melo, A.S., Fontana, L.F., da Silva, E., Bolanho, B.C., Moreno, B.P., Sarragiotto, M.H., Dias-Arieira. C.R. (2020). Aqueous extracts of *Crambe abyssinica* seed cake: Chemical composition and potential for nematode control. *Ind. Crops Prod.*, 156, 112860.

Thligene, N., Mezzapesa, G.N., Mondelli, D., Trani, A., Veronico, P., Melillo, M.T., Sasanelli, N. (2019). Effect of coffee silver skin and brewers' spent grain in the control of root-knot nematodes. *Helminthologia*, 56(1), 30–41.

Thongkaewyuana, A. and Chairin, T. (2018). Biocontrol of *Meloidogyne incognita* by *Metarhizium guizhouense* and its protease. *Biol. Control*, 126, 142–146.

Tseke, P.E. and Mashela P.W. (2017). Efficacy of Nemafric-BL phytonematicide from fresh *Cucumis africanus* fruit on suppression of root-knot nematodes and growth of tomato plants. *Res. Crops*, 18(2), 289–293.

Wang, C., Masler, E.P., Rogers, S.T. (2018). Responses of *Heterodera glycines* and *Meloidogyne incognita* infective juveniles to root tissues, root exudates, and root extracts from three plant species. *Plant Dis.*, 102(9), 1733–1740.

Wen, Y., Meyer, S.L.F., MacDonald, M.H., Zheng, L., Jing, C., Chitwood, D.J. (2019). Nematotoxicity of *Paeonia* spp. extracts and *Camellia oleifera* tea seed cake and extracts to *Heterodera glycines* and *Meloidogyne incognita*. *Plant Dis.*, 103, 2191–2198.

Zaidat, S.A.E., Mouhouche, F., Babaali, D., Abdessemed, N., De Cara, M., Hammache. M. (2020). Nematicidal activity of aqueous and organic extracts of local plants against *Meloidogyne incognita* (Kofoid and White) Chitwood in Algeria under laboratory and greenhouse conditions. *Egypt. J. Biol. Pest Control*, 30, 46.

5

Agro-industrial By-products and Waste as Sources of Biopesticides

Maria Fe ANDRES and Azucena GONZÁLEZ COLOMA
ICA, CSIC, Spain

5.1. Introduction

Chemical pesticides used in crop protection pose long-term threats and risks to living organisms due to their harmful side effects (Kumar et al. 2021). Over the past 50 years, crop protection has relied heavily on synthetic chemical pesticides, but their availability is now declining as a result of new health and safety legislation (Chandler et al. 2011; Andrés et al. 2018). Therefore, the development of green and environmentally safe pesticides (biopesticides) has become a great challenge. To search for natural product-based biopesticides, different sources can be used including plants, microorganisms, endophytes or agricultural waste. Agricultural residues are rich in bioactive compounds (Sadh et al. 2018) and can be used as an alternative source for the production of biopesticides.

Agro-industrial residues are materials derived from the processing of agroforestry industry products (Santana-Méridas et al. 2012). In recent years,

Biocontrol of Plant Disease,
coordinated by Claire PRIGENT-COMBARET and Bernard DUMAS. © ISTE Ltd 2022.

waste generation by agroforestry industries represented approximately 1.3 billion tons annually with a high cost of more than 1,000 billion dollars per year (FAO 2019; Esposito et al. 2020). Therefore, proper waste management is required to reduce the environmental impact of such waste.

Some of these residues can be used to generate bioenergy and are valorized by extracting bioactive products. Recent studies have shown that agro-industrial waste can be a potential natural source of antioxidants, antimicrobials (Shirahigue and Ceccato-Antonini 2020) and biopesticides (Santana-Méridas et al. 2012; Ntalli et al. 2020a; Sciubba et al. 2020).

There are numerous sources of agroforestry-derived wastes with reported biopesticidal effects, including residues from the fruit industry (citrus peel, cherimolla seeds, pomegranate peel, nuts), horticulture (potato), cereals, oilseed and fiber crops, conifer forestry, essential oil distillation, etc. (Bernardi et al. 2017; Benelli et al. 2018; Alhamda et al. 2019; Da Silva et al. 2019; Ribeiro et al. 2019; Wang et al. 2019; Aliaño-González et al. 2020; Huang et al. 2020; Ashamo et al. 2021; De Carvalho et al. 2021; Pacifico et al. 2021; Pavela et al. 2021; Farag et al. 2021a, 2021b, 2021c; Shea et al. 2021).

Among the largest amounts of waste are wood residues from forestry and wood-processing industries along with lignin-rich agricultural wastes. These raw materials need further processing (pyrolysis) to be biomass valorized, rich in bioactive natural products and derivatives (Mattos et al. 2019). Additionally, in Mediterranean countries, the olive oil industry produces enormous amounts of residues with detrimental and toxic effects on the environment (Abbattista et al. 2021). Also, in Mediterranean countries, the waste generated by the essential oil industry, at a smaller scale, represents an interesting source of bioactive natural products (Julio et al. 2017a, 2017b; Politi et al. 2020; Sharma and Kaur 2021).

This chapter focuses on the biopesticide potential of some selected processing-based wastes such as olive mill waste, essential oil-derived hydrolates and products from biomass pyrolysis against plant-parasitic nematodes and insect pests.

5.2. Biopesticidal properties of pyrolysis products from agro-industrial waste

In recent years, thermochemical conversion technologies such as pyrolysis have gained increasing interest in converting various types of waste biomass. Pyrolysis is an energy-efficient and robust process in an inert atmosphere at elevated temperatures (~300–700°C) that produces solid (biochar), liquid (bio-oil, vinegar) and gaseous end-products of varying quality and quantity from biomass materials (Hagner et al. 2020). The process results in waste reduction and can produce valuable products with biopesticide potential.

Biochar is a high-carbon solid product of pyrolysis with many environmental benefits. Incorporation of biochar made up of multiple feedstocks improved plant performance (Marra et al. 2018). Biochar application also has beneficial anti-nematode effects, although these effects varied in potency and mode of action (Poveda et al. 2021). However, there are few studies demonstrating the direct effects of biochar on plant-parasitic nematodes (Table 5.1). Biochar from burned wood added in coffee seedlings (greenhouse trial) was effective in suppressing the population of *Pratylenchus coffeae* (Rahayu and Sari 2017); water extracts from olive mill waste biochar produced at 300°C demonstrated a high nematicidal activity in vitro (on *Meloidogyne* incognita juveniles), but this activity was drastically reduced in biochars obtained at higher pyrolysis temperatures (Marra et al. 2018). Ikwunagu et al. (2019) reported that the reproductive capacity of *Meloidogyne* spp. was significantly affected by different biochar (bitter leaves, cassava peels and sawdust) applications in a pot experiment with *Vigna radiata*. Some biochars that are not direct nematicidal agents show a negative impact on plant-parasitic nematodes due to several mechanisms, including better solubilization and absorption of nutrients, which helps to improve plant growth and resistance to pathogen-related stress, or induced plant defense mechanisms (Elad et al. 2011). For example, a study in a microcosm trial demonstrated that wheat-straw biochar increased the abundance of soil fungi-feeding nematodes and decreased the plant-parasitic nematode populations (Zhang et al. 2013). On the other hand, the effective reduction of *M. incognita* population in rice by the addition of oak wood biochar to the potting medium was linked to a local H_2O_2 accumulation and a transcriptional enhancement of ethylene-related genes, which influenced plant defense genes

(Huang et al. 2015). Similarly, the application of biochar from coniferous wood and spelt husks decreased infection rates of the root-lesion nematode *P. penetrans* and associated biomass loss in carrots, the induced resistance being a possible explanation for the highly reduced infection (George et al. 2016). Even an inactive biochar from maize straw demonstrated synergistic and potentiating effects in furfural–biochar-based formulations against *M. incognita* in tomatoes (Abdelnabby et al. 2018). This interesting finding indicates that biochar could be a useful additive in the production and formulation of biopesticides. All these studies show a great variability in the biochar impact on plant-parasitic nematodes associated with the type of agro-industrial waste selected as feedstock and the pyrolysis conditions for the biochar manufacturing process.

The liquid product of pyrolysis is formed by one aqueous phase (also called pyroligneous extracts or wood vinegar), and another organic phase known as bio-oil (Bridgwater 2012). The main organic compounds are hydroxyl aldehydes, hydroxyl ketones, sugar, and acetic and carboxylic acids (Kim et al. 2020). The greatest potential of pyroligneous extracts and bio-oils in agriculture is their use as fertilizers and pesticides as part of sustainable agricultural practices (Pereira et al. 2021).

Some authors have pointed out the potential use of pyroligneous extracts from plant waste biomass as a promising source to control plant-parasitic nematodes (Farias et al. 2020). The use of pyroligneous extract from eucalyptus tar (Biopirol®) showed a high suppressive effect on population densities of *Pratylenchus* and *Meloidogyne* in sugarcane fields (Rossi and Lima 2007), and other studies demonstrated this efficacy in reducing egg hatching, second-stage juveniles and the reproduction rate of *M. incognita* and *M. javanica* in tomatoes (Corbani 2008; Melo et al. 2012) and lettuce (Santos et al. 2017). Recently, Farias et al. (2020) demonstrated the high in vitro nematicidal activities of pyroligneous extracts from the fruit endocarp (agro-industrial waste) of palm species against *Scutellonema bradys* and the effects of pyroligneous extract from *Cocos nucifera* palm trees on the reproduction rate of a mixed population of *S. bradys* and *Pratylenchus sp.* on yam plants in greenhouses (Farias et al. 2020).

Waste	Pyrolysis product	Application	Target species	References
Oak wood	Biochar	Addition in rice plants	*Meloidogyne incognita*	Huang et al. (2015)
Coniferous wood, spelt husks	Biochar	Addition in carrot plants	*Pratylenchus penetrans*	George (2016)
Burned log wood	Biochar	Water extracts (in vitro bioassays) Addition in coffee plants	*P. coffeae*	Rahayu and Sari (2017)
Olive mill	Biochar	Water extracts (in vitro bioassays)	*M. incognita*	Marra et al. (2018)
Bitter leaves, cassava peel and sawdust	Biochar	Addition in *Vigna radiata* plant	*Meloidogyne sp.*	Ikwunagu et al. (2019)
Eucalyptus tar	Pyroligneous extracts	Addition in sugarcane fields Water extracts (in vitro bioassays) Addition in tomato plants Addition in lettuce plants	*Meloidogyne sp.*, *Pratylenchus sp.* *M. incognita*, *M. javanica*	Rossi and Lima, (2007) Corbani (2008) and Melo et al. (2012) Santos et al. (2017)
Fruit endocarp of palm trees species *Cocos nucifera* palm tree	Pyroligneous extracts	Water extracts (in vitro bioassays) Addition in yam plants	*Scutellonema bradys* *S. bradys*, *Pratylenchus sp.*)	Farias et al. (2020)

Table 5.1. *Nematicidal pyrolysis products from different agro-industrial wastes*

Some pyrolysis-derived bio-oils from different raw materials have also been reported as being insecticidal. Table 5.2 shows the residues, the main components identified and target insects.

The main components present in the bioactive bio-oils depend on the starting raw material, and can be affected by the pyrolysis process (fast, slow, fraction used, etc.). The volatile fraction obtained from the pyrolysis of wood components (cellulose, lignin) along with alfalfa and maize residues have been tested against *Bactrocera oleae*, an important olive pest (Bonanomi et al. 2021). Insecticidal polycyclic aromatic hydrocarbons were the main components of lignin, cellulose and hemicellulose bio-oil (Hossain et al. 2013, 2015). Acetic acid, 2-propanone, acetol, 2-furancarboxaldehyde and phenolic compounds were the main components of the lignocellulose bio-oil obtained from *Vitex pubescens*, with anti-termite effects (Oramahi et al. 2013, 2014). Vanillin, p-cresol, 2,6-dimethoxyphenol and 2,6-dimethylphenol were identified in the anti-termite bio-oil from *Dyera costulata* wood (Kadir et al. 2021). Other lignin-rich residues explored for the production of insecticidal bio-oils include coconut shells with insecticidal, and anti-termitic effects (Wititsiri 2011) and oil palm trunk, rich in acetol, acetic acid and phenolic derivatives and also active on termites (Oramahi et al. 2018).

Residues other than wood-related ones have been explored for bio-oil production. Tobacco leaves transformed into bio-oil gave different fractions, some rich in nicotine and others in phenolic compounds, active on *Aedes aegypti* (Jufri et al. 2016) and *Leptinotarsa decemlineata* (Booker et al. 2010a, 2010b; Hossain et al. 2016). Spent coffee grains gave bio-oil fractions active on *L. decemlineata* (Bedmutha et al. 2011; Hossain et al. 2016). Bioactive phenolic compounds (Hossain et al. 2016) and other components such as neophytadiene, phytol and fatty acids (Cáceres, et al. 2015) active on *L. decemlineata* were obtained from tomato plant bio-oil. Canola and mustard leaves gave bio-oil fractions rich in hexadecanoic and octadecanoic fatty acids and bioactive on *L. decemlineata* (Suqi et al. 2014). Hagner et al. (2020) showed that the slow pyrolysis liquid of willow, with a high content of acetic acid and phenolic compounds, was a very effective insecticide against the oat aphid, *Rhopalosiphum padi*. Bamboo vinegar, mainly composed of phenol derivatives and acetic acid, was insecticidal to the fruit fly *Drosophila melanogaster* (Alias et al. 2019). A recent study of Urrutia et al. (2021) reported the insecticidal efficacy of bio-oil obtained from acid pretreated sunflower seed hulls against stored-product pests such as *Sitophilus oryzae*, *Lasioderma serricorne* and *Tribolium castaneum*.

Material	Waste	Pyrolysis product	Main components	Target species	References
Wood and related	Wood, alfalfa, maize	Volatile fraction as smoke water		*Bactrocera oleae*	Bonanomi et al. (2021)
	Wood components	Bio-oil (550°C)	Polycyclic aromatic hydrocarbons	*Leptinotarse decemlineata*, *Acyrthosiphon pisum*	Hossain et al. (2013, 2015)
	Vitex pubescens	Bio-oil (350, 400, 450°C) fractions	Acetic acid, 2-propanone, acetol, 2-furancarboxaldehyde, phenolic compounds (2-methoxyphenol, 2,6- dimethoxyphenol)	*Coptotermes curvignathus*, *C. formosanus*, *R. speratus*	Oramahi et al. (2013, 2014)
	Dyera costulata	Bio-oil, (500°C)	Vanillin, p-cresol, 2,6-dimethoxyphenol, 2,6-dimethylphenol	*C. curvignathus*	Kadir et al. (2021)
	Coconut shell	Bio-oil (300–400°C) fractions		*Ferrisia virgata*, *Odontotermes sp.*	Wititsiri (2011)
	Palm trunk (*Elaeis sp.*)	Bio-oil (350, 400, 450°C) fractions	Acetol, acetic acid and phenolic derivatives	*Coptotermes formosanus*	Oramahi et al. (2018)
	Willow	Bio-oil (280°C)	Acetic acid and phenolic compounds	*Rhopalosiphum padi*	Hagner et al. (2020)

Material	Waste	Pyrolysis product	Main components	Target species	References
Plant and seed residues	Tobacco leaves	Bio-oil (500°C)	Nicotine, limonene, indole, pyridine	*Aedes aegypti*	Jufri et al. (2016)
		Bio-oil (350, 400, 450, 500, 550, 600°C) fractions	Alkaloidal (nicotine) and phenolic fractions	*Leptinotarsa decemlineata*	Booker et al. (2010a, 2010b)
	Tobacco leaves, tomato plant, spent coffee	Dried 210°C bio-oil (300–400°C 2D MFR reactor) fractions	Phenolic compounds	*L. decemlineata*	Hossain et al. (2016)
	Tomato dry residue	Bio-oil (300 and 500°C) fractions	Neophytadiene, phytol, fatty acids	*L. decemlineata*	Cáceres et al. (2015)
	Coffee grains	Bio-oil (400 and 550°C) fractions	Phenolic compounds, others	*L. decemlineata*	Bedmutha et al. (2011)
	Canola and mustard leaves	Bio-oil (300 and 500°C) fractions	Hexadecanoic, octadecanoic fatty acids	*L. decemlineata*	Suqi et al. (2014)
	Bamboo	Vinegar	Phenol derivatives and acetic acid	*Drosophila melanogaster*	Alias et al. (2019)
	Sunflower seed hulls	Acid pretreated hulls bio-oil		*Sitophilus oryzae*, *Lasioderma serricorne*, *Tribolium castaneum*	Urrutia et al. (2021)

Table 5.2. *Relevant insecticidal pyrolysis bio-oils from different agro-industrial wastes*

All of these studies indicate that pyrolysis liquids and bio-oils are a promising source of bioactive ingredients for integrated pest management. As mentioned before, the selection of the agro-industrial waste as raw materials and the pyrolysis technology used will determine the quality and their potential to be commercialized (Hagner et al. 2020). Further investigations on the active components and modes of action of these pyrolysis products are needed.

5.3. Biopesticidal properties of hydrolates: by-products of essential oil distillation

Essential oils are obtained from the distillation process of aromatic plants, while the plant material after the distillation (solid residues), as well as the hydrolates or hydrosols (water residues from the distillation), are by-products of such process. In Europe, the perfumes and fragrances industry and the demand for organic ingredients in personal care is expected to positively influence the growth of the hydrosols market (https://www.marketresearchfuture.com/reports/hydrosols-market-4789). Their global economic impact is projected to grow at a rate of 5.17% between 2019 and 2024, and reach a market value of USD 437 million by the end of 2024 (https://www.marketresearchfuture.com/reports/hydrosols-market-4789).

These aromatic waters are used as food preservatives (Sagdic et al. 2013; Ozturk 2015; D'Amato, et al. 2018; Tanhaş et al. 2020; Xiao et al. 2020), cosmetic and perfumery (Kurkcuoglu and Baser 2003) and aromatherapy (Catty 2001) and have reported antimicrobial and antioxidant effects (Wilkinson et al. 2003; Njimoh et al. 2015; Prusinowska et al. 2016; Verma et al. 2016; Allal, et al. 2019; Georgiev et al. 2019; Jakubczyk et al. 2021; Kaddour et al. 2021).

In agriculture, the antifungal effects of hydrosols against plant pathogens have been extensively described (Wilkinson et al. 2003; Boyraz and Özcan 2005, 2006; Özcan 2005; Moon et al. 2007; Inouye et al. 2009; Tabti et al. 2014; Belabbes et al. 2017; Tabet Zatla et al. 2017; Sebaa et al. 2019; Zatla et al. 2020; Paramalingam et al. 2021).

Apart from the potential application of hydrolates as antifungal agents, these residues have reported nematicidal effects (Table 5.3). Julio et al. (2017b) reported the nematicidal properties of the hydrolate of the *Artemisia absinthium* variety (Candial®) against *M. javanica* juveniles, egg hatching, juvenile root penetration and their strong effects on nematode reproduction traits in potted tomato plants for the first time. The active compounds were present in the hydrolate organic fraction, characterized by (5Z)-2,6-dimethylocta-5,7-dien-2,3-

diol as the main compound. Nematicidal effects in vitro and in vivo of hydrolates from Spanish aromatic plant species *L.* × *intermedia* var. super, *L. luisieri*, *T. vulgaris* and *T. zygis* have been proved against the root-knot nematode *M. javanica*. Hydrolate treatments caused a high mortality of juveniles, inhibited egg hatching, suppressed juvenile root penetration and reduced the nematode infection of tomato plants. The hydrolates from the *Thymus* species concentrated the active compounds in the organic fraction, with thymol being the major component (Andres et al. 2018). Furthermore, hydrosols from Greek aromatic species *Origanum vulgare, Mentha piperita* and *Melissa officinalis* were evaluated against *M. javanica* and *M. incognita. Origanum vulgare* hydrosol, mainly composed of carvacrol and thymol, showed in vitro high activity on both nematode species (Ntalli et al. 2020b). Recently, *Satureja hellenica* and *Cuminum cyminum* hydrosols have showed in vitro nematicidal effects on motility, hatching and survival in soil of second-stage juveniles of *M. javanica* and *M. incognita*. The nematicidal activity of *S. hellenica* hydrosol could be attributed to major components, carvacrol and thymol, and *Cuminum cyminum* hydrosol exhibited y-terpinen-7-al, cumin aldehyde and α terpinen-7-al as its main compounds (Pavardella et al. 2020, 2021).

Hydrosols/hydrolates also have potential applications against insect pests (Table 5.4). For example, hydrosols from *Eucalyptus globulus* have been reported as active on mealybugs (Sharma and Kaur 2021); *Lavandula* × intermedia on *Tribolium confusum* (Politi et al. 2020); *Ocimum basilicum* and *Ruta chalepensis* on *Aphis gossypii* and *Tetranychus urticae* (Traka et al. 2018) and *Mentha suaveolens* and *M. pulegium* on *Toxoptera aurantia* (Zekri et al. 2016). Improved extraction methods such as microwave-assisted extraction of *Origanum majorana, Mentha pulegium* and *Melissa officinalis* resulted in shorted distillation times and less energy consumption and gave hydrosols with similar composition and effects on *Myzus persicae* to those obtained by conventional hydrodistillation (Petrakis et al. 2015).

Many of the reports on the biocidal (nematicidal and insecticidal) effects of hydrosols/hydrolates tested the aqueous extracts obtained by conventional hydrodistillation (laboratory distillation on a Clevenger apparatus) without further processing, making these extracts difficult to standardize and replicate. Furthermore, there are only three reports on hydrolates further extracted to give organic extracts that have been tested and chemically characterized by means of bioguided chromatography (Julio et al. 2017a, 2017b; Andres et al. 2018). The extraction of the organic fraction of the hydrolate makes possible to standardize and characterize (chemical markers, chemical fingerprinting) different batches of hydrolate.

Source species	Residue	Extract type	Main components	Target species	References
Artemisia absinthium variety (Candial®)	Hydrolate (vapor-pressure pilot plant extraction)	Ethanolic extract of hydrolate-adsorbed active charcoal	(5Z)-2,6-dimethylocta-5,7-diene-2,3-diol	*Meloidogyne javanica*	Julio et al. (2017)
Lavandula × *intermedia*	Hydrolate (vapor-pressure pilot plant extraction)	Organic and water fractions (DCM)		*M. javanica*	Andres et al. (2018)
L. luisieri					
Tymus. vulgaris			Thymol		
T. zygis			Thymol		
Origanum vulgare	Hydrosol (conventional hydrodistillation)	Water extract	Carvacrol and thymol	*M. incognita* *M. javanica*	Ntalli et al. (2020)
Satureja hellenica	Hydrosol (conventional hydrodistillation)	Water extract	Carvacrol and thymol	*M. incognita* *M. javanica*	Pavardella et al. (2020)
Cuminum cyminum	Hydrosol (conventional hydrodistillation)	Water extract	γ-terpinen-7-al, cumin aldehyde and α terpinen-7-al	*M. incognita* *M. javanica*	Pavardella et al. (2021)

Table 5.3. *Nematicidal activity of hydrolate/hydrosol by-products from essential oil distillation*

Source species	Residue	Extract type	Main components	Target species	References
Eucalyptus globulus	Hydrosol (conventional hydrodistillation)	Water extract	Phenolics and tannins	Mealybugs (family Pseudococcidae)	Sharma and Kaur (2021)
Lavandula × intermedia	Hydrosol (conventional hydrodistillation)	Water extract	1,8-Cineole, linalool, camphor	*Tribolium confusum*	Politi et al. (2020)
Mentha suaveolens,	Hydrosol (conventional hydrodistillation,)	Water extract	Piperitenone oxide, l	*Toxoptera aurantii*	Zekri et al. (2016)
M. pulegium			Carvacrol		
Ocimum basilicum	Hydrosol (conventional hydrodistillation)	Water extract	Linalool, eugenol, cineol	*Aphis gossypii, Tetranychus urticae*	Traka et al. (2018)
Ruta chalepensis			2-Nonanone, 2-undecanone		
Origanum majorana	Hydrosol (conventional and microwave-assisted hydrodistillation)	Water extract	Terpinen-4-ol	*Myzus persicae*	Petrakis et al. (2015)
Mentha pulegium			Piperitone		
Melissa officinalis			Neral, geranial		
Lavandula luisieri	Hydrolate (vapor-pressure pilot plant extraction)	Ethanolic extract of hydrolate-adsorbed active charcoal	3,3,4,5-Tetramethyl-2H-pyran-2,6(3H)-dione, (2,2,3,4-tetramethyl-5-oxocyclopent-3-en-1-yl)-methyl acetate	*Hyalomma lusitanicum*	Julio et al. (2017)

Table 5.4. *Insecticidal activity of hydrolate/hydrosol by-products from essential oil distillation*

5.4. Biopesticidal properties of olive oil mill waste

The production of high-quality olive oils implies the generation of vast quantities of solid residues and/or wastewater that may have a great impact on terrestrial and aquatic environments because of their high phytotoxicity. Disposal and/or treatment of olive mill wastewater or semisolid waste are among the most important environmental problems in the Mediterranean region due to the presence of phenols, lipids and fatty acids (Pantziaros et al. 2021).

When two-phase decanters are used for horizontal centrifugation, a humid semi-solid pomace is obtained as a by-product (alperujo or olive mill waste OP). In the case of the three-phase centrifugal technology, solid pomace (OP) and olive oil mill wastewater (OMW) or alpechin are the two major by-products (Abbatista et al. 2021). Olive pomace and olive oil mill wastewater are included in the agricultural livestock waste category (Decision 2000/532/EC of the European Union Commission) (European Environment Agency 2009).

OP is mostly used for the recovery of oil by solid–liquid extraction with hexane. After extraction, OP can be applied on the soil as a herbicide and compost (Tortosa et al. 2018). Due to its polysaccharide content, OP can be a source of gelling pectic material (Lama-Muñoz et al. 2012). OP can also be used as a substrate for the biotechnological fermentation of useful microorganisms to produce high value-added products (Ghilardi et al. 2020). OP is rich in phenolic compounds and triterpenic acids, with numerous biological activities, including anti-inflammatory, antitumor, antimicrobial, antioxidant, antidiabetic and cardio-protective activities (Medina et al. 2018). One of the most interesting bioactive phenols that can be recovered from OP is hydroxytyrosol, currently used as a therapeutic agent, dietary supplement or natural ingredient in food and feed industries (Nunes et al. 2016).

OMWW (alpechin) consists mainly of water with 3–15% organic matter. Phenolic compounds, sugars and organic acids are the main components of the organic fraction, and it is considered the most polluting waste generated by the agri-food industries (Khdair and Abu-Rumman 2020). OMWW can be very harmful to soil microflora, plants and freshwater species given its high concentrations of phenols and flavonoids with strong antimicrobial and phytotoxic properties (Justino et al. 2012; Leris et al. 2019). The potential application of OMWW phenolic compounds for the preparation of functional beverages and enriched olive oils (Caporaso et al. 2018) and their potential for weed control (Tubeileh et al. 2019) has been explored.

OMW contains compounds with nematicidal action (Table 5.5) and has been suggested for use as a biopesticide (Leontopoulos et al. 2020). Phenolic compounds contained in these wastes could play an important role among phytochemical compounds in nematode management (Chitwood 2002). Vouyoukalou and Stefanoudaki (1998) demonstrated in vitro and in vivo nematicidal effects of OMW extracts on *Meloidogyne incognita*. Egg hatch was significantly reduced by the ethyl acetate extract, and OMW water caused a significant reduction of galling and egg masses on tomato roots. Additionally, phenolic compounds such as tyrosol, caffeic acid, benzoic acid, coumaric acid and vanillic acid isolated from these active OMW extracts showed nematicidal activity. Water extracts of two-phase OMW extracts strongly affected the second-stage juvenile (J2) motility and inhibited egg hatch of *M. incognita* (Cayuela et al. 2008). Similarly, Marra et al. (2018) reported strong in vitro activity on *M. incognita* juveniles of an OMW water extract mainly composed of fatty acids, phenolic compounds and triterpene acids.

Source species	Residue	Extract type	Main components	Target species	References
Olea europea	Olive mill wastewater	Aqueous ethyl acetate	Tyrosol, caffeic acid, benzoic acid, coumaric acid and vanillic acid	*Meloidogyne incognita*	Vouyoukalou and Stefanoudaki (1998)
	Olive mill wastewater	Aqueous		*M. incognita*	Cayuela et al. (2008)
	Olive mill wastewater	Aqueous	Fatty acids, phenolic compounds and triterpene acids	*M. incognita*	Marra et al. (2018)

Table 5.5. *Nematicidal activity of olive oil mill wastewater*

The insecticidal effects of olive leaf extracts on several insect pests such as *Dacus oleae* (Lo Scalzo et al. 1994), *Schistocerca gregaria* (Barbouche et al. 1996), *Euphyllura olivina* (Ouguas et al. 2010), *Lymantria dispar* (Hemming and Lindroth 2000) and *Acyrthosiphon pisum* (Golawska et al. 2008) have been attributed to their content in phenolic and polyphenolic compounds (Lo Scalzo et al. 1994). However, the insecticidal potential of olive mil waste phenolic fractions has been poorly explored (Table 5.6).

Polyphenolic extracts from OMW have been tested as biopesticides against the Mediterranean fruit fly *Ceratitis capitata* (Di Ilio and Cristofaro 2021); crude OMW was larvicidal to *Potosia opaca* var. cardui Gyllenhal (Boutaj et al. 2020). Phenolic OMW extracts caused mortality to *Euphyllura olivina* and *Aphis citricola* (Larif et al. 2013), and a hydroxytyrosol-rich preparation prepared from fresh OMW using a hydrolysis and post-hydrolysis purification process was an effective insecticide against the olive psyllid, *Euphyllura olivina* (Debo et al. 2011).

Source species	Residue	Extract type	Main components	Target species	References
Olea europea	Olive mill wastewater	Aqueous	Polyphenolic fractions	*Ceratitis capitata*	Di Ilio and Cristofaro (2021)
	Olive mill wastewater	Aqueous	Monomeric phenols: hydroxytyrosol, tyrosol	*Potosia opaca var. cardui Gyllenhal*	Boutaj et al. (2020)
	Olive mill wastewater	Aqueous	Phenolic extracts	*Euphyllura olivina, Aphis citricola*	Larif et al. (2013)
	Olive mill wastewater	Aqueous processed (hydrolysis and purification)	Hydroxytyrosol	*Euphyllura olivina*	Debo et al. (2011)

Table 5.6. *Insecticidal activity of olive oil mill wastewater*

Active compounds present in OMW represent a promising option for the control of pests and diseases. However, the biopesticide potential of these waste materials has been barely investigated, probably due to its undesired phytotoxic effects. Further research is needed, focusing on OMW processing by the microbial transformations and the selective extraction of the bioactive components to reduce their phytotoxicity increasing their value (Leontopoulos et al. 2020).

5.5. Conclusion and future directives

Agricultural residues rich in bioactive compounds can be used as an alternative source for the production of biopesticides. Among the largest amount of waste is wood residue from forestry. Additionally, in Mediterranean countries, the olive oil industry produces residue and the waste generated by the essential oil industry are also significant sources of bioactive natural products.

Wood and lignin-rich residue can be processed by pyrolysis that produces solid (biochar), liquid (bio-oil, vinegar) and gaseous end-products of varying quality and quantity from biomass materials. The process results in waste reduction and can produce valuable products with biopesticide potential.

Direct effects of biochar on plant-parasitic nematodes have been demonstrated. Pyroligneous extracts (liquids) have also reported effects on plant-parasitic nematodes and insects, with acetic acid, phenolic compounds and fatty acids being the main components of these active fractions. All of these studies indicate that pyrolysis biochar, liquids and bio-oils are a promising source of bioactive ingredients for integrated pest management. The selection of the agro-industrial waste as raw materials and the pyrolysis technology used will determine the quality and their potential to be commercialized. Further investigations on the active components and modes of action of these pyrolysis products are needed.

The production of essential oils from aromatic plants generates plant material (solid residues) and hydrolates or hydrosols (water residues from the distillation) as by-products. In agriculture, the antifungal effects of hydrosols against plant pathogens have been extensively described. Additionally, these residues have nematicidal and insecticidal effects. Most studies on biocidal effects of hydrosols/hydrolates tested the aqueous extracts obtained by conventional hydrodistillation (laboratory distillation on a Clevenger apparatus) without further processing, making these extracts difficult to standardize and replicate. The extraction of the organic fraction of the hydrolate makes possible to standardize and characterize (chemical marker, chemical fingerprinting) different batches of hydrolates.

The production of high-quality olive oils implies the generation of vast quantities of solid residues (OP) and/or wastewater (OMW) that may have a great impact on terrestrial and aquatic environments. Many compounds present in OMW and OMWW represent a promising option for the control of pests and diseases in the areas where olive oil production occurs. OP can be applied on the soil as herbicide, and it is rich in phenolic compounds and triterpenic acids. OMWW (alpechin), the most polluting waste generated by the agri-food industry, also contains phenolic compounds, sugars and organic acids and has been evaluated for direct weed control. OMW has nematicidal action. However, the insecticidal potential of olive mil waste phenolic fractions has been poorly explored. These olive waste materials have undesired phytotoxic effects. However, further waste processing (biotransformation, selective extractions, etc.) could reduce the phytotoxicity, increasing their value. Therefore, more research is needed, focusing on microbial transformations of olive mill waste and the selective extraction of the bioactive components as a potential source of new safe biopesticides.

5.6. Acknowledgments

This work was supported by grants LIFE 2017 ENV/ES/000192 (WASTE4GREEN) and Ministerio de economía y competitividad, PID2019-106222RB-C31/SRA (State Research Agency, 10.13039/501100011033).

5.7. References

Abbattista, R., Ventura, G., Calvano, C.D., Cataldi, T.R.I., Losito, I. (2021). Bioactive compounds in waste by-products from olive oil production: Applications and structural characterization by mass spectrometry techniques. *Foods*, 10(6), 1236.

Abdelnabby, H., Hu, Z., Wang, H. (2018). Furfural–biochar-based formulations show synergistic and potentiating effects against *Meloidogyne incognita* in tomato. *J. Pest. Sci.*, 91, 203–218.

Alhamda, S., Barlian, E., Razak, A., Sari, M., Herawati, N. (2019). Effectiveness of soursop seed extract (*Annona muricata* L) as a natural repellent on *Aedes aegypti*. *IJITEE*, 8(12), 5012–5018.

Aliaño-González, M.J., Richard, T., Cantos-Villar, E. (2020). Grapevine cane extracts: Raw plant material, extraction methods, quantification, and applications. *Biomolecules*, 10(8), 1–33.

Alias, N.Z., Abdullah, S., Shaari, S.S., Junik, J.J., Kamal, M.L., Masdar, N.D. (2019). The potential of bamboo vinegar (*Gigantochloa albociliata*) as insecticide. In *Charting the Sustainable Future of ASEAN in Science and Technology*, Alias, N.Z. and Yusof, R. (eds). Springer, Singapore.

Allal, A., Bellifa, S., Benmansour, N., Selles, C., Semaoui, M., Hassaine, H., Muselli, A. (2019). Essential oil and hydrosol extract chemical profile, antioxidant and antimicrobial potential of *Daphne gnidium* L. from Algeria. *J. Essent. Oil-Bear. Plants*, 22(5), 1277–1288.

Andrés, M.F., González-Coloma, A., Muñoz, R., De la Peña, F., Julio, L.F., Burillo, J. (2018). Nematicidal potential of hydrolates from the semi industrial vapor-pressure extraction of Spanish aromatic plants. *Env. Sci. Pollut. Res.*, 25(30), 29834–29840.

Ashamo, M.O., Ileke, K.D., Ogungbite, O.C. (2021). Entomotoxicity of some agro-wastes against cowpea bruchid, *Callosobruchus maculatus* (Fab.) [Coleoptera: Chrysomelidae] infesting cowpea seeds in storage. *Heliyon*, 7(6), e07202.

Barbouche N., Dhouib S., Ammar M., Ben Hamouda M.H. (1996). Action de trois substrats alimentaires : Bersim, faux poivrier et olivier sur le développement de la cuticule chez le criquet pèlerin *Schistocerca gregaria*. *Annales de l'INRAT*, 69, 131–146

Bedmutha, R., Booker, C.J., Ferrante, L., Briens, C., Berruti, F., Yeung, K.K., Scott, I., Conn, K. (2011). Insecticidal and bactericidal characteristics of the bio-oil from the fast pyrolysis of coffee grounds. *J. Anal. Appl. Pyrolysis*., 90(2), 224–231.

Belabbes, R., Dib, M.E.A., Djabou, N., Ilias, F., Tabti, B., Costa, J., Muselli, A. (2017). Chemical variability, antioxidant and antifungal activities of essential oils and hydrosol extract of *Calendula arvensis* L. from Western Algeria. *Chem. Biodiversity*, 14, e1600482.

Benelli, G., Pavela, R., Lupidi, G., Nabissi, M., Petrelli, R., Ngahang Kamte, S.L., Cappellacci, L., Fiorini, D., Sut, S., DalL'Acqua, S. et al. (2018). The crop-residue of fiber hemp cv. Futura 75: From a waste product to a source of botanical insecticides. *Env. Sci. Pollut. Res.*, 25(11), 10515–10525.

Bernardi, D., Ribeiro, L., Andreazza, F., Neitzke, C., Oliveira, E.E., Botton, M., Nava, D.E., Vendramim, J.D. (2017). Potential use of *Annona* by products to control *Drosophila suzukii* and toxicity to its parasitoid *Trichopria anastrephae*. *Ind. Crop. Prod.*, 110, 30–35.

Bonanomi, G., Jesu, G., Zotti, M., Idbella, M., D'Errico, G., Laudonia, S., Vinale, F., Abd-Elgawad, A. (2021). Biochar-derived smoke-water exerts biological effects on nematodes, insects, and higher plants but not fungi. *Sci. Total Environ.*, 750, 142307.

Booker, C.J., Bedmutha, R., Scott, I.M., Conn, K., Berruti, F., Briens, C., Yeung, K.K. (2010a). Bioenergy II: Characterization of the pesticide properties of tobacco bio-oil. *Int. J. Chem. React. Eng.*, 8, Article A26.

Booker, C.J., Bedmutha, R., Vogel, T., Gloor, A., Xu, R., Ferrante, L., Yeung, K.K., Scott, I.M., Conn, K.L., Berruti, F. et al. (2010b). Experimental investigations into the insecticidal, fungicidal, and bactericidal properties of pyrolysis bio-oil from tobacco leaves using a fluidized bed pilot plant. *Ind. Eng. Chem. Res.*, 49(20), 10074–10079.

Boutaj, H., Boutasknit, A., Anli, M., Ait Ahmed, M., El Abbassi, A., Meddich, A. (2020). Insecticidal effect of olive mill wastewaters on *Potosia opaca* (Coleoptera: Scarabeidae) larva. *Waste Biomass Valorization*, 11(7), 3397–3405.

Boyraz, N. and Özcan, M. (2005). Antifungal effect of some spice hydrosols. *Fitoterapia*, 76(7–8), 661–665.

Boyraz, N. and Özcan, M. (2006). Inhibition of phytopathogenic fungi by essential oil, hydrosol, ground material and extract of summer savory (*Satureja hortensis* L.) growing wild in Turkey. *Int. J. Food Microbiol.*, 107(3), 238–242.

Bridgwater, A.V. (2012). Review of fast pyrolysis of biomass and product upgrading. *Biomass Bioenergy*, 38, 68e94.

Cáceres, L.A., McGarvey, B.D., Briens, C., Berruti, F., Yeung, K.K., Scott, I.M. (2015). Insecticidal properties of pyrolysis bio-oil from greenhouse tomato residue biomass. *J. Anal. Appl. Pyrolysis.*, 112, 333–340.

Caporaso, N., Formisano, D., Genovese, A. (2018). Use of phenolic compounds from olive mill wastewater as valuable ingredients for functional foods. *Crit. Rev. Food Sci. Nutr.*, 58(16), 2829–284.

Catty, S. (2001). *Hydrosols: The Next Aromatherapy*. Healing Arts Press, Rochester, VT.

Cayuela, M.L., Millner, P.D., Meyer, S.L., Roig, A. (2008). Potential of olive mill waste and compost as biobased pesticides against weeds, fungi, and nematodes. *Sci. Total Environ.*, 399, 11–18.

Chandler, D., Bailey, A.S., Tatchell, G.M., Davidson, G., Greaves, J., Grant, W.P. (2011). The development, regulation and use of biopesticides for integrated pest management. *Philos. Trans. R. Soc. Lond. B: Biol. Sci.*, 366, 1987–1998.

Chitwood, D.J. (2002). Phytochemical based strategies for nematode control. *Annu. Rev. Phytopathol.*, 40, 221–249.

Corbani, R.Z. (2008). Estudo do extrato pirolenhoso Biopirol® no manejo de nematoides em cana-de-açúcar, olerícolas e citros, em diferentes ambientes. Doctoral thesis, São Paulo State University, São Paulo.

D'Amato, S., Serio, A., López, C.C., Paparella, A. (2018). Hydrosols: Biological activity and potential as antimicrobials for food applications. *Food Control*, 86(1), 126–137.

Da Silva, T.C., De Camargo, A.B., Sanches, L.A., Garlet, J. (2019). *Musa acuminata* pseudostem extract on the control of *Atta sexdens rubropilosa*. *Biosci. J.*, 35(2), 459–466.

De Carvalho, S.S., Ribeiro, L.P., Forim, M.R., Da Silva, M.F.G.F., Bicalho, K.U., Fernandes, J.B., Vendramim, J.D. (2021). Avocado kernels, an industrial residue: A source of compounds with insecticidal activity against silverleaf whitefly. *Environ. Sci. Pollut. Res.*, 28(2), 2260–2268.

Debo, A., Yangui, T., Dhouib, A., Ksantini, M., Sayadi, S. (2011). Efficacy of a hydroxytyrosol-rich preparation from olive mill wastewater for control of olive psyllid, *Euphyllura olivina*, infestations. *Crop Prot.*, 30(12), 1529–1534.

Di Ilio, V. and Cristofaro, M. (2021). Polyphenolic extracts from the olive mill wastewater as a source of biopesticides and their effects on the life cycle of the Mediterranean fruit fly *Ceratitis capitata* (Diptera, Tephriditae). *Int. J. Trop. Insect Sci.*, 41(1), 359–366.

Elad, Y., Cytryn, E., Harel, Y.M., Lew, B., Graber, E.R. (2011). The biochar effect: Plant resistance to biotic stresses. *Phytopathol. Mediterr.*, 50, 335–349.

Esposito, B., Sessa, M.R., Sica, D., Malandrino, O. (2020). Towards circular economy in the agri-food sector. A systematic literature review. *Sustainability*, 12, 7401.

European Environment Agency (2009). Diverting waste from landfill: Effectiveness of waste management policies in the European Union. *EEA Rep.*, 1–68.

FAO (2019). The state of food and agriculture 2019: Moving forward on food loss and waste reduction [Online]. Available at: http://www.fao.org/3/ca5162en/ca5162en.pdf.

Farag, S.M., Essa, E.E., Alharbi, S.A., Alfarraj, S., Abu El-Hassan, G.M.M. (2021a). Agro-waste derived compounds (flax and black seed peels): Toxicological effect against the West Nile virus vector, *Culex pipiens* L. with special reference to GC–MS analysis. *Saudi J. Biol. Sci.*, 28(9), 5261–5267.

Farag, S.M., Hussein, M.A., Hafez, S.E., Khaled, A.S., Kamel, O.M., Zyaan, O.H. (2021b). Larvicidal, biological, and histopathological alterations induced by pomegranate peel extract, *Punica granatum*, against *Culex pipiens* I. (Diptera: Culicidae). *Egypt. J. Aquat. Biol. Fish.*, 25(2), 139–161.

Farag, S.M., Kamel, O.M., Abu El-Hassan, G.M.M., Zyaan, O.H. (2021c). Larvicidal and repellent potential of *Sesamum indicum* hull peels extracts against *Culex pipiens* l. (Diptera: Culicidae). Egypt. *J. Aquat. Biol. Fish.*, 25(2), 995–1011.

Farias, S.P., Almeida, A.V., Nascimento, E.S., Soletti, J., Balliano, T.L., Filho, G.M., Muniz, M. (2020). In vitro and in vivo control of yam dry rot nematodes using pyroligneous extracts from palm trees. *Ceres*, 67(6), 482–490.

George, C., Kohler, J., Rillig, M.C. (2016). Biochars reduce infection rates of the root-lesion nematode *Pratylenchus penetrans* and associated biomass loss in carrot. *Soil Biol. Biochem.*, 95, 11–18.

Georgiev, V., Ananga, A., Dincheva, I., Badjakov, I., Gochev, V., Tsolova, V. (2019). Chemical composition, in vitro antioxidant potential, and antimicrobial activities of essential oils and hydrosols from Native American muscadine grapes. *Molecules*, 24(18), 3355.

Ghilardi, C., Sanmartin Negrete, P., Carelli, A.A., Borroni, V. (2020). Evaluation of olive mill waste as substrate for carotenoid production by *Rhodotorula mucilaginosa*. *Bioresour. Bioprocess.*, 7(52).

Golawska, S., Kapusta, I., Lukasik, I., Wojcicka, A. (2008). Effect of phenolicson the pea aphid, *Acyrthosiphon pisum* (Harris) population on *Pisum sativum* L (Fabaceae). *Pestycydy*, (3–4), 71–77.

Hagner, M., Tiilikkala, K., Lindqvist, I., Niemelä, K., Wikberg, H., Källi, A., Rasa, K. (2020). Performance of liquids from slow pyrolysis and hydrothermal carbonization in plant protection. *Waste Biomass Valorization*, 11(3), 1005–1016.

Hemming, J.D.C. and Lindroth, R.L. (2000). Effects of phenolic glycosides and protein on gypsy moth (Lepidoptera: Lymantriidae) and forest tent caterpillar (Lepidoptera: Lasiocampidae) performance and detoxication activities. *Environ. Entomol.*, 29(6), 1108–1115.

Hossain, M.M., Scott, I.M., McGarvey, B.D., Conn, K., Ferrant E.L., Berruti, F., Briens, C. (2013). Toxicity of lignin, cellulose and hemicellulose-pyrolyzed bio-oil combinations: Estimating pesticide resources. *J. Anal. Appl. Pyrolysis*, 99, 211–216.

Hossain, M.M., Scott, I.M., McGarvey, B.D., Conn, K., Ferrante, L., Berruti, F., Briens, C. (2015). Insecticidal and anti-microbial activity of bio-oil derived from fast pyrolysis of lignin, cellulose, and hemicellulose. *J. Pest Sci.*, 88(1), 171–179.

Hossain, M.M., Scott, I.M., Berruti, F., Briens, C. (2016). Application of 1D and 2D MFR reactor technology for the isolation of insecticidal and anti-microbial properties from pyrolysis bio-oils. *J. Environ. Sci. Health B*, 51(12), 860–867.

Huang, W.K., Ji, H.L., Gheysen, G., Debode, J., Kyndt, T. (2015). Biochar-amended potting medium reduces the susceptibility of rice to root-knot nematode infections. *BMC Plant Biol.*, 15, 267.

Huang, X., Li, T., Shan, X., Lu, R., Hao, M., Lv, M., Sun, Z., Xu, H. (2020). High value-added use of citrus industrial wastes in agriculture: Semisynthesis and anti-tobacco mosaic virus/insecticidal activities of ester derivatives of limonin modified in the B ring. *J. Agric. Food Chem.*, 68(44), 12241–12251.

Hydrosols Market (2020). Research report – Global forecast til 2024. Market Research Future® [Online]. Available at: https://www.marketresearchfuture.com /reports/hydrosols-market-4789 [Accessed 3 September 2021].

Ikwunagu, E.A., Ononuju, C.C., Orikara, C.C. (2019). Nematicidal effects of different biochar sources on root-knot nematode (*Meloidogyne* spp.) egg hatchability and control on mungbean (*Vigna radiata* (l.) Wilczek). *Int. J. Entomol. Nematol.*, 4(2), 1–14.

Inouye, S., Takahashi, M., Abe, S. (2009). Inhibitory activity of hydrosols, herbal teas and related essential oils against filament formation and the growth of Candida albicans. *Nihon Ishinkin Gakkai Zasshi*, 50(4), 243–252.

Jakubczyk, K., Tuchowska, A., Janda-Milczarek, K. (2021). Plant hydrolates – Antioxidant properties, chemical composition and potential applications. *Biomed. Pharmacother.*, 142, 112033

Jufri, M., Irmayanti, E., Gozan, M. (2016). Formulation of tobacco based mosquito repellent to avoid dengue fever. *Int. J. Pharmtech Res.*, 9(7), 140–145.

Julio, L.F., Barrero, A.F., Herrador Del Pino, M.M., Arteaga, J.F., Burillo, J., Andres, M.F., Díaz, C.E., González-Coloma, A. (2016). Phytotoxic and nematicidal components of *Lavandula luisieri*. *J. Nat. Prod.*, 79(2), 261–266.

Julio, L.F., Díaz, C.E., Aissani, N., Valcarcel, F., Burillo, J., Olmeda, S., González-Coloma, A. (2017a). Ixodicidal compounds from pre-domesticated *Lavandula luisieri*. *Ind. Crop. Prod.*, 110, 83–87.

Julio, L.F., González-Coloma, A., Burillo, J., Diaz, C.E., Andrés, M.F. (2017b). Nematicidal activity of the hydrolate by-product from the semi industrial vapor pressure extraction of domesticated *Artemisia absinthium* against *Meloidogyne javanica*. *Crop. Prot.*, 94, 33–37.

Justino, C.I.L., Pereira, R., Freitas, A.C., Rocha-Santos, T.A.P., Panteleitchouk, T.S.L., Duarte, A.C. (2012). Olive oil mill wastewaters before and after treatment: A critical review from the ecotoxicological point of view. *Ecotoxicology*, 21(2), 615–629.

Kaddour, F., Aissaoui, N., Dib, M.E.A., Bensaid, O., Muselli, A. (2021). Chemical composition and antimicrobial activity of essential oil and hydrosol extract from roots of *Plumbago europaea* and in-vitro combinatory antimicrobial effect of hydrosol extract with gentamicin and amphotericin b. *Nat. Prod. J.*, 11(2), 169–179.

Kadir, R., Sarif Mohd Ali, M., Kartal, S.N., Elham, P., Mohd Ali, N.A., Awang, A.F. (2021). Chemical characterization of pyrolysis liquids from *Dyera costulata* and evaluation of their bio-efficiency against subterranean termites, *Coptotermes curvignathus*. *Eur. J. Wood and Wood Prod.* [Online]. Available at: https://doi.org/10.1007/s00107-021-01732-z.

Khdair, A. and Abu-Rumman, G. (2020). Sustainable environmental management and valorization options for olive mill by-products in the Middle East and North Africa (MENA) region. *Processes*, 8(6), 671.

Kumar, J., Ramlal, A., Mallick, D., Mishra, V. (2021). An overview of some biopesticides and their importance in plant protection for commercial acceptance. *Plants*, 10, 1185.

Kurkcuoglu, M. and Baser, K.H.C. (2003). Studies on Turkish rose concrete, absolute, and hydrosol. *Chem. Nat. Compd.*, 39(5), 457–464.

Lama-Muñoz, A., Rodríguez-Gutiérrez, G., Rubio-Senent, F., Fernández-Bolaños, J. (2012). Production, characterization and isolation of neutral and pectic oligosaccharides with low molecular weights from olive by-products thermally treated. *Food Hydrocoll.*, 28(1), 92–104.

Larif, M., Zarrouk, A., Soulaymani, A., Elmidaoui, A. (2013). New innovation in order to recover the polyphenols of olive mill wastewater extracts for use as a biopesticide against the *Euphyllura olivina* and *Aphis citricola*. *Res. Chem. Intermed.*, 39(9), 4303–4313

Leontopoulos, S., Skenderidis, P., Vagelas, I.K. (2020). Potential use of polyphenolic compounds obtained from olive mill waste waters on plant pathogens and plant parasitic nematodes. In *Plant Defence: Biological Control. Progress in Biological Control*, Mérillon, J.M. and Ramawat, K.G. (eds). Springer, Cham.

Leris, I., Kalogianni, E., Tsangaris, C., Smeti, E., Laschou, S., Anastasopoulou, E., Vardakas, L., Kapakos, Y., Skoulikidis, N.T. (2019). Acute and sub-chronic toxicity bioassays of olive mill wastewater on the eastern mosquitofish *Gambusia holbrooki*. *Ecotoxicol. Environ. Saf.*, 175, 48–57.

Lo Scalzo, R., Scarpati, M.L., Verzegnassi, B., Vita, G. (1994). *Olea europaea* chemical repellent to *Dacus oleae* females. *J. Chem. Ecol.*, 20(8), 1813–1923.

Marra, R., Vinale, F., Cesarano, G., Lombardi, N., D'Errico, G., Crasto, A., Mazzei, P., Piccolo, A., Incerti, G., Woo, S.L. et al. (2018). Biochars from olive mill waste have contrasting effects on plants, fungi and phytoparasitic nematodes. *PLoS One*, 13(6), e0198728.

Mattos, C., Veloso, M.C.C., Romeiro, G.A., Folly, E. (2019). Biocidal applications trends of bio-oils from pyrolysis: Characterization of several conditions and biomass, a review. *J. Anal. Appl. Pyrolysis*, 139, 1–12.

Medina, E., Romero, C., Brenes, M. (2018). Residual olive paste as a source of phenolic compounds and triterpenic acids. *Eur. J. Lipid Sci. Technol.*, 120(4), 1438–7697.

Melo, T.A., Serra, I.M.R.S., Silva, G.S., Sousa, R.M.S. (2012). Produtos naturais aplicados para manejo de *Meloidogyne incognita* em tomateiros. *Summa Phytopathologica*, 38, 223–227.

Moon, T., Cavanagh, H.M.A., Wilkinson, J.M. (2007). Antifungal activity of australian grown *Lavandula* spp. essential oils against *Aspergillus nidulans*, *Trichophyton mentagrophytes*, *Leptosphaeria maculans* and *Sclerotinia sclerotiorum*. *J. Essent. Oil Res.*, 19(2), 171–175.

Njimoh, D.L., Assob, J.C.N., Mokake, S.E., Nyhalah, D.J., Yinda, C.K., Sandjon, B. (2015). Antimicrobial activities of a plethora of medicinal plant extracts and hydrolates against human pathogens and their potential to reverse antibiotic resistance. *Int. J. Microbiol.*, 547156.

Ntalli, N., Adamski, Z., Doula, M., Monokrousos, N. (2020a). Nematicidal amendments and soil remediation. *Plants*, 9, 429.

Ntalli, N., Ozalexandridou, E.X., Kasiotis, K.M., Samara, M., Golfinopoulos, S.K. (2020b). Nematicidal activity and phytochemistry of Greek *Lamiaceae* species. *Agronomy*, 10(8), 1119.

Nunes, M.A., Pimentel, F.B., Costa, A.S.G., Alves, R.C., Oliveira, M.B.P.P. (2016). Olive by-products for functional and food applications: Challenging opportunities to face environmental constraints. *Innov. Food Sci. Emerg. Technol.*, 35, 139–148.

Oramahi, H.A. and Nurhaida F.D. (2014). New bio preservatives from lignocelluloses biomass bio-oil for anti-termites *Coptotermes curvignathus* Holmgren. *Procedia Environ. Sci.*, 20, 778–784.

Oramahi, H.A. and Yoshimura, T. (2013). Antifungal and antitermitic activities of wood vinegar from *Vitex pubescens* Vahl. *J. Wood Sci.*, 59, 344–350.

Oramahi H.A., Yoshimura, T., Diba, F., Setyawati, D. (2018). Antifungal and antitermitic activities of wood vinegar from oil palm trunk. *J. Wood Sci.*, 64, 311–317.

Ouguas, Y., Hilal, A., Elhadram, I. (2010). Effet biocide des extraits phénoliques oléicoles sur les adultes du psylle de l'olivier *Euphyllura olivinacosta* (Homoptera: Psyllidae) sur deux variétés d'oliviers, Menara et Arbequine, au Maroc. *Revue Ezzaitouna*, 11(1), 1–15.

Özcan, M. (2005). Effect of spice hydrosols on the growth of *Aspergillus parasiticus* NRRL 2999 strain. *J. Med. Food*, 8(2), 275–278.

Ozturk, I. (2015). Antifungal activity of propolis, thyme essential oil and hydrosol on natural mycobiota of Sucuk, a Turkish fermented sausage: Monitoring of their effects on microbiological, color and aroma properties. *J. Food Process. Preserv.*, 39(6), 1148–1158.

Pacifico, D., Lanzanova, C., Pagnotta, E., Bassolino, L., Mastrangelo, A.M., Marone, D., Matteo, R., Lo Scalzo, R., Balconi, C. (2021). Sustainable use of bioactive compounds from *Solanum tuberosum* and Brassicaceae wastes and by-products for crop protection: A review. *Molecules*, 26(8), 2174.

Pantziaros, A.G., Trachili, X.A., Zentelis, A.D., Sygouni, V., Paraskeva, C.A. (2021). A new olive oil production scheme with almost zero wastes. *Biomass Convers. Biorefin.*, 11(2), 547–557.

Paramalingam, P., Anuar, M.S.K., Baharum, N.A., Abdullah, J.O., Aziz, J.A., Saidi, N.B. (2021). In vitro evaluation of antifungal activity of selected Malaysian plants against the wilt pathogen of banana, *Fusarium oxysporum* f.sp. *cubense* tropical race 4. *Malaysian J. Sci.*, 40(2), 16–24.

Pardavella, I., Nasiou, E., Daferera, D., Trigas, P., Giannakou, I. (2020). The use of essential oil and hydrosol extracted from *Satureja hellenica* for the control of *Meloidogyne incognita* and *M. javanica*. *Plants*, 9(7), 1–13.

Pardavella, I., Daferera, D., Tselios, T., Skiada, P., Giannakou, I. (2021). The use of essential oil and hydrosol extracted from *Cuminum cyminum* seeds for the control of *Meloidogyne incognita* and *Meloidogyne javanica*. *Plants*, 10(1), 46.

Pavela, R., Maggi, F., Mazzara, E., Torresi, J., Cianfaglione, K., Benelli, G., Canale, A. (2021). Prolonged sublethal effects of essential oils from non-wood parts of nine conifers on key insect pests and vectors. *Ind. Crop. Prod.*, 168, 113590.

Pereira, E.G., Fauller, H., Magalhães, M., Guirardi, B., Martins, M.A. (2021). Potential use of wood pyrolysis coproducts: A review. *Environ. Prog. Sustainable Energy*, 41(1), e13705.

Petrakis, E.A., Kimbaris, A.C., Lykouressis, D.P., Polissiou, M.G., Perdikis, D.C. (2015). Hydrosols evaluation in pest control: Insecticidal and settling inhibition potential against *Myzus persicae* (Sulzer). *J. Appl. Entomol.*, 139(4), 260–267.

Politi, M., Menghini, L., Conti, B., Bedini, S., Farina, P., Cioni, P.L., Braca, A., De Leo, M. (2020). Reconsidering hydrosols as main products of aromatic plants manufactory: The lavandin (*Lavandula × intermedia*) case study in Tuscany. *Molecules*, 25(9), 2225.

Poveda, J., Martínez Gómez, Á., Fenoll, C., Escobar, C. (2021). The use of biochar for plant-pathogen control. *Phytopathology*, E33529050.

Prusinowska, R., Śmigielski, K., Stobiecka, A., Kunicka-Styczyńska, A. (2016). Hydrolates from lavender (*Lavandula angustifolia*): Their chemical composition as well as aromatic, antimicrobial and antioxidant properties. *Nat. Prod. Res.*, 30(4), 386–393.

Rahayu, D.S. and Sari, N.P. (2017). Development of pratylenchus coffeae in biochar applied soil, coffee roots and its effect on plant growth. *Pelita Perkebunan*, 33, 24–32.

Ribeiro, N.C., Da Camara, C.A.G., Melo, J.P.R., De Moraes, M.M. (2019). Acaricidal properties of essential oils from agro-industrial waste products from citric fruit against *Tetranychus urticae. J. Appl. Entomol.*, 143(7), 731–743.

Rossi, C.E. and Lima, C.B. (2007). Controle alternativo de nematoides em cultura orgânica de cana-de-açúcar. *Revista Brasileira de Agroecologia*, 2, 1545–1548.

Sadh, P.K., Duhan, S., Duhan, J.S. (2018). Agro-industrial wastes and their utilization using solid state fermentation: A review. *Bioresour. Bioprocess*, 5, 1.

Sagdic, O., Ozturk, I., Tornuk, F. (2013). Inactivation of non-toxigenic and toxigenic *Escherichia coli* O157:H7 inoculated on minimally processed tomatoes and cucumbers: Utilization of hydrosols of *Lamiaceae* spices as natural food sanitizers. *Food Control.*, 30, 7–14.

Santana-Méridas, O., González-Coloma, A., Sánchez-Vioque, R. (2012). Agricultural residues as a source of bioactive natural products. *Phytochem. Rev.*, 11, 447–466.

Santos, A.M., de Almeida, F.A., Fonseca, W.L., Leite, M.L.T., Pereira, F.F., Carvalho, R.M. (2017). Ácido pirolenhoso no manejo de nematoides das galhas na cultura da alface. *Espacios*, 38(43), 1–9.

Sciubba, F., Chronopoulou, L., Pizzichini, D., Lionetti, V., Fontana, C., Aromolo, R., Socciarelli, S., Gambelli, L., Bartolacci, B., Finotti, E. et al. (2020). Olive mill wastes: A source of bioactive molecules for plant growth and protection against pathogens. *Biology*, 9, 450.

Sebaa, N.A., Zatla, A.T., Dib, M.E.A., Tabti, B., Costa, J., Muselli, A. (2019). Antifungal activity of essential oil and hydrosol extract of *Ballota nigra* L. and their protective effects against the black rot of tomatoes. *Curr. Nutr. Food Sci.*, 15(7), 662–671.

Sharma, A.D. and Kaur, I. (2021). By-product hydrosol of *Eucalyptus globulus* essential oil distillation as source of botanical insecticides: Wealth from waste. *Not. Sci. Biol.*, 13(1), 10854.

Shea, E., Fernandez-Bayo, J.D., Pastrana, A.M., Simmons, C.W. (2021). Identification and evaluation of volatile organic compounds evolved during solarization with almond hull and shell amendments. *J. Air Waste Manag. Assoc.*, 71(3), 400–412.

Shirahigue, L.D. and Ceccato-Antonini, S.R. (2020). Agro-industrial wastes as sources of bioactive compounds for food and fermentation industries. *FOOD TECHNOLOGY Cienc. Rural*, 50(4), e20190857.

Suqi, L., Caceres, L., Schieck, K., Booker, C.J., McGarvey, B.M., Yeung, K.K., Pariente, S., Briens, C., Berruti, F., Scott, I.M. (2014). Insecticidal activity of bio-oil from the pyrolysis of straw from *Brassica* spp. *J. Agric. Food Chem.*, 62(16), 3610–3618.

Tabet Zatla, A., Dib, M.E.A., Djabou, N., Ilias, F., Costa, J., Muselli, A. (2017). Antifungal activities of essential oils and hydrosol extracts of *Daucus carota* subsp. sativus for the control of fungal pathogens, in particular gray rot of strawberry during storage. *J. Essent. Oil Res.*, 29(5), 391–399.

Tabti, L., El Amine Dib, M., Djabou, N., Benyelles, N.G., Paolini, J., Costa, J., Muselli, A. (2014). Control of fungal pathogens of *Citrus sinensis* L. by essential oil and hydrosol of *Thymus capitatus* L. *J. Appl. Bot. Food Qual.*, 87, 279–285.

Tanhaş, E., Martin, E., Korucu, E.N., Dirmenci, T. (2020). Effect of aqueous extract, hydrosol, and essential oil forms of some endemic *Origanum* L. (*Lamiaceae*) taxa on polyphenol oxidase activity in fresh-cut mushroom samples. *J. Food Process. Preserv.*, 44(10), e14726.

Tortosa, G., González-Gordo, S., Ruiz, C., Bedmar, E.J., Palma, J.M. (2018). "Alperujo" compost improves the ascorbate (Vitamin C) content in pepper (*Capsicum annuum* L.) fruits and influences their oxidative metabolism. *Agronomy*, 8(6), 82.

Traka, C.K., Petrakis, E.A., Kimbaris, A.C., Polissiou, M.G., Perdikis, D.C. (2018). Effects of *Ocimum basilicum* and *Ruta chalepensis* hydrosols on *Aphis gossypii* and *Tetranychus urticae*. *J. Appl. Entomol.*, 142(4), 413–420.

Tubeileh, A.M., Schnorf, J.T., Mondragon, I., Gray, G.A. (2019). Exploiting olive mill byproducts and other waste for organic weed management. *Horticulturae*, 5(3), 59.

Urrutia, R.I., Yeguerman, C., Jesser, E., Gutierrez, V.S., Volpe, M.A., Werdin González, J.O. (2021). Sunflower seed hulls waste as a novel source of insecticidal product: Pyrolysis bio-oil bioactivity on insect pests of stored grains and products. *J. Clean. Prod.*, 287, 125000.

Verma, R.S., Padalia, R.C., Saikia, D., Chauhan, A., Krishna, V., Sundaresan, V. (2016). Chemical composition and antimicrobial activity of the essential oils isolated from the herbage and aqueous distillates of two *Thymus* species. *J. Essent. Oil-Bear. Plants*, 19(4), 936–943.

Vouyoukalou, E. and Stefanoudaki, E. (1998). Nematicidal activity of waste water from olive oil-mills. *Nemat. Medit.*, 26, 157–160.

Wang, Z., Cáceres, L.A., Hossain, M.M., Abdallah, S.B., Ogbeide, O., Yao, Z., Renaud, J.B., Scott, I.M. (2019). The Antioxidant and enzyme inhibitory activity of balsam fir (*Abies balsamea* (L.) Mill.) bark solvent extracts and pyrolysis oil. *Waste Biomass Valorization*, 10(11), 3295–3306.

Wilkinson, J.M., Hipwell, M., Ryan, T., Cavanagh, H.M.A. (2003). Bioactivity of *Backhousia citriodora*: Antibacterial and antifungal activity. *J. Agric. Food Chem.*, 51(1), 76–81.

Wititsiri, S. (2011). Production of wood vinegars from coconut shells and additional materials for control of termite workers, *Odontotermes* sp. and striped mealy bugs, *Ferrisia virgata*, Songklanakarin. *J. Sci. Technol.*, 33(3), 349–354.

Xiao, Y., He, J., Zeng, J., Yuan, X., Zhang, Z., Wang, B. (2020). Application of citronella and rose hydrosols reduced enzymatic browning of fresh-cut taro. *J. Food Biochem.*, 44(8), e13283.

Zatla, A.T., Mami, I., Dib, M.E.A., Sifi, M.E.A. (2020). Efficacy of essential oil and hydrosol extract of *Marrubium vulgare* on fungi responsible for apples rot. *Anti-infect. Agents*, 18(3), 285–293.

Zekri, N., Handaq, N., El Caidi, A., Zair, T., Alaoui El Belghiti, M. (2016). Insecticidal effect of *Mentha pulegium* L. and *Mentha suaveolens* Ehrh. hydrosols against a pest of citrus, *Toxoptera aurantii* (Aphididae). *Res. Chem. Intermed.*, 42(3), 1639–1649.

Zhang, X.Q.L., Wenju, L., Zhang, M., Bao, X., Xie, Z. (2013). Soil nematode response to biochar addition in a Chinese wheat field. *Pedosphere*, 23, 98–103.

Zulkefli, N.F., Ghani, W.A., Ali, S., Asib, N., Chowmasundaram, Y. (2021). Development of bio-pesticides from bio-oil of oil palm biomass waste (Palm kernel shell) against *Metisa plana* Walker bagworm (Lepidoptera: Psychidae). *Food Res.*, 5, 137–143.

6

Antimicrobial and Defense Elicitor Peptides as Biopesticides for Plant Disease Control

Emilio MONTESINOS, Esther BADOSA, Maria PLA, Laura MONTESINOS and Anna BONATERRA
INTEA, University of Girona, Spain

There is a need for novel pesticides in plant disease protection. Functional peptides are candidates to biopesticides because they fulfil the requirements of sustainability and originate from living organisms. Their function is mainly antimicrobial and is involved in the regulation of the immune response of animals and plants. Peptides from microorganisms, mainly bacteriocins, cyclic lipopeptides and peptaibols, are notably antimicrobial and play a key role in the biocontrol of plant bacterial and fungal pathogens. Peptides in plants have a role in their immune response, either as part of the damage perception system via the pattern triggered immunity (PTI) or as released products with antimicrobial action. In animals, peptides also have a role in defense as the first barrier of the immune system due to their antimicrobial activity or as immunomodulators. Several peptides from animals and plants have been used as native or synthetic derivatives against plant pathogens, either applied topically or as heterologous expression in plants. Synthetic peptides, either de novo or obtained by optimizing natural sequences, have been developed and engineered to improve their properties or for plant expression. The

Biocontrol of Plant Disease,
coordinated by Claire PRIGENT-COMBARET and Bernard DUMAS. © ISTE Ltd 2022.

production of several functional peptides intended for plant disease protection has been achieved using microbial and plant biofactories. However, there are still challenges for their use in plant disease protection related to the regulatory framework, the high cost of production, the need for formulation and proper delivery systems.

6.1. Introduction

Approximately 33% of crop productivity of the main six crops in agriculture is lost due to pests and diseases, despite the measures to counteract their adverse effects (Oerke 2006). Most countries have undergone a drastic reduction of the authorized pesticides in order to minimize their non-target effects in the environment and consumers. Consequently, several pests and diseases may be insufficiently or non-controlled, especially taking into account the emergence of resistance within the plant pathogen populations. Consequently, there is a need for novel pesticides, with minimum non-target effects and sustainable production under the scenario of climate change, and biopesticides are an alternative or complement to conventional pesticides. The need for novel pesticides is especially important in the case of bacterial disease control, and more necessary against emerging diseases of economic importance (list including Xf/HLB, etc.), as well as the increasing reports of resistance to existing pesticide compounds in fungal and bacterial plant pathogens. In addition, novel fungicides and bactericides should fit into the current rules for biopesticides (EU, USA regulations) that must be sustainable, biodegradable, environmentally compatible and mainly based on natural/biological structures. Functional peptides fit into the scope of biopesticides and are candidates to novel compounds for plant protection (Montesinos 2007; Marcos 2008; Rajaseharan et al. 2013). Peptides are considered polypeptides of less than 50 aa, mostly produced by living organisms, either by ribosomal protein synthesis or by NRPS, involved in antagonism/antibiosis in microorganisms and immune defense, or stress mitigation in animals and plants.

Peptides are classified according to several criteria that include their activity, origin, structure and amino acid composition (Rahnamaeian et al. 2017; Kumar et al. 2018). The most frequent classification is based on the structure they can adopt in vivo like α-helix, β sheet, β-hairpins or looped configuration and linear peptides with unusual bias (see Figure 6.1). Peptides adopting α-helical structure are commonly found in animals and are generally potent antimicrobials that are cationic and amphipathic. Other peptides commonly in plants have β sheet structure and contain disulfide bonds between conserved cysteines or are cyclic. Several peptides have β-hairpins or looped configurations like some bacteriocin lantibiotics. The last group are linear peptides, generally rich in amino acids like Pro, Gly, His or Trp.

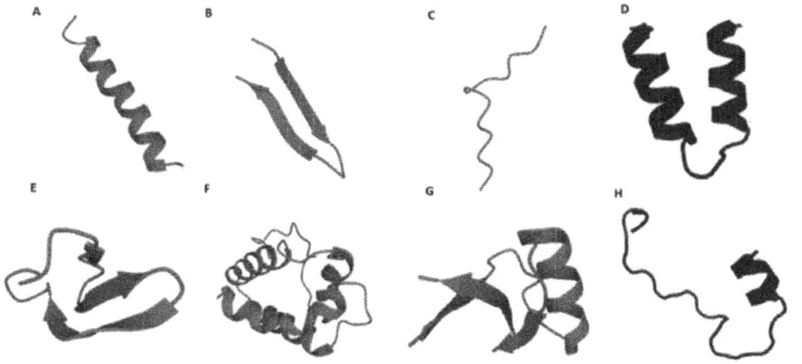

Figure 6.1. *Structures of antimicrobial and plant defense elicitor peptides. A, magainin; B, tachyplesin; C, indolicidin; D, synthetic peptide BP178; E, kalata cyclotide; F, LTP from plants; G, plant defensin; H, Rosaceae consensus PEP peptide. The structures were made according to their aminoacid sequences using the Protein Data Bank (PDB) (www.rcsb.org) or PEP-FOLD3 (https://bioserv.rpbs.univ-paris-diderot.fr/services/PEP-FOLD3/). For a color version of this figure, see www.iste.co.uk/prigentcombaret/biocontrol.zip*

The most known group of functional peptides exhibits antimicrobial activity due to cell membrane or intracellular damage in the target microorganism that results in its lysis or inactivation. Most lytic peptides are cationic amphipathic and interact with the negatively charged phospholipids and LPS of the bacterial membrane, mainly in Gram negative bacteria, through a postulated toroidal, barrel-shape or carpet-like mechanism, resulting in pores and membrane destabilization (Beckinger 2004). However, lysis is not only restricted to peptides that adopt an alpha-helix structure because it is also observed in beta-sheet or linear peptides. A group of interesting peptides can penetrate the target cell without significant damage to the membrane, but may provoke an adverse effect on protein or nucleic acid synthesis or cell division, or may inhibit proteinases. Another type of functional peptides is able to prime plants by inducing plant defense responses against pathogens, mitigate abiotic stress or stimulate their growth (Lori et al. 2015). Another group of interesting peptides is "cell penetrating" with no antimicrobial activity or priming effect on plants that can be used to deliver cargo molecules to the cytosol (Eggenberger et al. 2011).

Developing peptides for plant protection requires a long process that is presented in Figure 6.2. The process usually starts by finding natural sources of peptides that are used to build up a peptide library or as templates for developing synthetic analogs. These peptide libraries are screened in vitro and in planta against plant pathogens of interest, and the selected candidates are further studied and produced, by synthesis or eventually in biofactories, for later formulation as biopesticides.

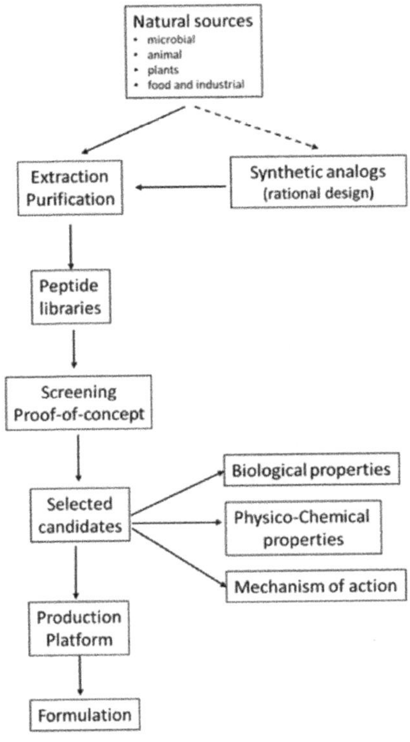

Figure 6.2. *Workflow of the discovery and development of functional peptides for plant protection. Natural sources of peptides are used to build up a peptide library or selected members used as models for developing synthetic analogs. The peptide libraries are screened against plant patho-systems of interest, and the lead candidates are studied in further detail and produced for later formulation as biopesticides*

6.2. Peptides of microbial origin

Microorganisms synthesize different bioactive peptides that exhibit a wide range of relevant activities such as antimicrobial, cytotoxic, antitumor, host cell immune eliciting and surfactant activities. The main types of bioactive peptides produced by microorganisms are bacteriocins, cyclic lipopeptides, peptaibols and pseudopeptides (see Table 6.1).

Type	Peptide	Microorganism
Bacteriocins	Agrocin	*Agrobacterium radiobacter*
	Amylocyclicin	*Bacillus amyloliquefaciens*
	Carocin	*Pectobacterium carotovorum*
	Ericin	*Bacillus amyloliquefaciens*
	Glicinecin	*Xanthomonas axonopodis* pv. glycines
	Putidacin	*Pseudomonas putida*
	Pyocin	*P. fluorescens*
	Serracin	*Serratia plymuthica*
	Tailocin	*P. fluorescens*
	Thuricin	*Bacillus thuringiensis*
CLPs	Iturins, fengicins, surfactins	*Bacillus* spp.
	Chromobactomycin	*Chromobacterium* spp.
	Massetolide	*Pseudomonas fluorescens*
	Nunamycin	*P. fluorescens*
	Orfamide	*P. protegens*
	Poaemide	*P. poae*
	Syringomycin	*P. syringae*
	Syringopeptin	*P. syringae*
	Tensin	*P. fluorescens*
Peptaibols	Atroviridin	*Trichoderma atroviride*
	Harzianins	*T. harzianum*
	Peptaivirins	*Trichoderma* sp.
	Trichokonins	*T. pseudokoningii*
	Trichorzins	*T. harzianum*
Other (pseudopeptides, depsipeptides)	Chaetomin	*Chaetomium globosum*
	Coronamycin	*Streptomyces* spp.
	Cyclothiazomycin	*Streptomyces* spp.
	Glomecidin	*S. lavendulae*
	Neopeptins	*Streptomyces* sp.
	Pantocins	*Pantoea agglomerans*
	Polyoxins	*Streptomyces cacaoi*
	Syringolin	*P. syringae* pv. syringae
	Valinomycin	*Serratia plymuthica*

Compiled from: Yun et al. (2000), Lavermicocca et al. (2002), Ongena et al. (2005), Stein (2005), Kim et al. (2007), Romero et al. (2007), Luo et al. (2010), Raaijmakers et al. (2010), Abriouel et al. (2011), Scholz et al. (2014), Mora et al. (2015), Ma et al. (2017), Principe et al. (2018), Sarwar et al. (2018), Osada (2019), Smits et al. (2019), Khen et al. (2020) and Zhang et al. (2020).

Table 6.1. *Antimicrobial peptides produced by strains of bacteria and fungi that are active against plant pathogenic fungi and bacteria*

6.2.1. *Bacteriocins*

Bacteriocins are a heterogeneous group of ribosomal synthesized peptides produced by bacteria that have bacteriostatic or bactericidal activity against other related and unrelated microorganisms (Abriouel et al. 2011; Subramanian and Smith 2015). Bacteriocins generally exert their antimicrobial action by membrane pore formation, nonspecific degradation of cellular DNA, cleavage of 16S rRNA or tRNA, or inhibition of peptidoglycan synthesis resulting in cell lysis. The host strain is protected from its own bacteriocin through interaction with the immunity protein that is coproduced with the bacteriocin. Bacteriocins are classified into different classes, depending on their size, stability, translational modifications and presence of specific amino acids (e.g. lanthionine).

Several bacteriocins are active or mediate in the mechanism of action of biocontrol agents like Bac-GM17 from *Bacillus clausii* strain GM17 that have activity against *Agrobacterium tumefaciens* (Mouloud et al. 2013) or amylocyclicin produced by *B. amyloliquefaciens* FZB42 that have activity against *Ralstonia solanaceraum* and *Xanthomonas campestris* (Scholz et al. 2014). In addition, ericin S obtained from *B. subtilis* A1/3 have antibacterial activity against *Clavibacter michiganensis* in tomato plants (Stein et al. 2002) and thuricin Bn1 from *Bacillus thuringiensis* subsp. *kurstaki* Bn1 have activity against *P. savastanoi* and *P. syringae* (Ugras et al. 2013).

Bacteriocins from Gram-negative bacteria and strains producing them have also been studied against plant pathogens. Agrocin 84 produced by *Agrobacterium radiobacter* K84 is active against *Agrobacterium tumefaciens*, the causal agent of crown gall disease (Kim et al. 2006), or Serracin P produced by *Serratia plymithicum* J7 that is active against *Erwinia amylovora* (Jabrane et al. 2002). In addition, *Pseudomonas* spp. produces different active bacteriocins such as pyocin S or tailocin SF4c with activity against *Pseudomonas syringae* and *Xanthomonas* spp. pathogens (Principe et al. 2018) or a bacteriocin produced by *P. syringae* pv. *ciccaronei* NCPPB2355 active against *P. syringae* subsp. *savastanoi*, the causal agent of olive knot disease (Lavermicocca et al. 2002).

6.2.2. *Cyclic lipopeptides*

Cyclic lipopeptides (CLPs) are nonribosomal synthesized amphiphilic compounds, composed of a fatty acid tail linked to a short oligopeptide, which form a macrocyclic ring structure. In cyclodepsipeptides (CLDPs) one or more of its amide (peptide bond) groups are replaced by an ester bond.

They are predominantly produced by *Bacillus*, *Paenibacillus* and *Pseudomonas*, and act either directly, exhibiting broad-spectrum antimicrobial activities by targeting membranes, forming pores and leading to an imbalance in transmembrane ion fluxes, or indirectly as elicitors of plant defense in the host plant (Raaijmakers et al. 2010; Caulier et al. 2019).

Bacillus species synthesize a wide range of CLPs that have a cyclic structure comprising β-hydroxy or β-amino fatty acid chains incorporated through peptide bonds. They are grouped into iturins, fengycins and surfactins (Stein 2005). The iturins structure consists of cyclic heptapeptides (e.g. iturin A, iturin C, iturin D, iturin E, bacillomycin D, bacillomycin F, bacillomycin Lc, mojaversin A, mycosubtilin or mixirin), fengycins and plipastatins are decapeptides and surfactins are heptapeptides.

Pseudomonads produce CLPs containing chains of 7–25 aminoacids of which several form a lactone ring. They can be divided into several groups based on their structural relationships: viscosins, syringomycins, syringopeptins, amphisins, putisolvins and tolaasins (Geudens and Martins 2018). Many of the CLPs are biosurfactants and are a common feature of both plant-beneficial and pathogenic (Flury et al. 2017).

There are several examples of *Bacillus* spp. strains producing CLPs, responsible for the antifungal activity that protects plants from diseases. The fengycin produced by *Bacillus subtilis* NCD-2 plays a major role in the biocontrol of cotton seedling damping-off disease (Guo et al. 2014); fengycin and bacillomycin produced by *B. subtilis* SQR 9 is responsible for the suppression of Fusarium wilt of cucumber (Cao et al. 2012); bacillomycin, fengycin and iturin A produced by *Bacillus subtilis* are key factors in the antagonism against *Podosphaera fusca* (Romero et al. 2007). Some *Bacillus* strains producing CLPs also have antibacterial activity like *Bacillus* spp. SS12.9 that produce iturin and showed activity against *Xanthomonas oryzae* (Beric et al. 2012) or *Bacillus amyloliquefaciens* A17 that produce bacillomycin, fengycin, iturin and surfactin, which act synergistically against several bacterial plant pathogens (Mora et al. 2015). Surfactin produced by *Bacillus amyloliquefaciens* KPS46 is required to suppress infections by *Xanthomonas axonopodis* pv. glycines (Preecha et al. 2010) and in *Bacillus amyloliquefaciens* SQR9 contributed to the antagonism against *Sclerotinia sclerotium*, *Rhizoctonia solani* and *Fusarium solani* (Li et al. 2014a). Moreover, the purified surfactin A obtained from *Bacillus* NH-100 and NH-217 had strong antifungal activity against *Fusarium oxysporum*, *F. moniliforme*, *F. solani*, *Trichoderma atroviride* and *T. reesei* (Sarwar et al. 2018). Several CLPs produced by *Bacillus* spp. triggered defense responses in plants against pathogenic fungi such as fengycins and surfactins that induced significant protection in bean and tomato leaves against the fungal pathogen *Botrytis cinerea* (Ongena et al. 2007).

In the case of CLPs produced by *Pseudomonas*, they play a role in diverse bacterial functions such as growth-inhibitory activities against a broad range of microorganisms (Flury et al. 2017). For example, syringomycin E and syringopeptin 25A produced by *Pseudomonas syringae pv. syringae* B359 have a broad-range antifungal activity (Fogliano et al. 2002); tensin produced by *Pseudomonas fluorescens* 96.578 (Nielsen et al. 2000) or poaeamide produced by *Pseudomonas poae* RE1-1-14 have antifungal activity against *Rhizoctonia solani* (Zachow et al. 2015). Some CLPs produced by Pseudomonads are active, inducing defense-related responses in plants, such as massetolide belonging to the viscosin family that inhibit *Phytophthora infestans* by inducing local and systemic resistance (Tran et al. 2007) or orfamide A synthesized by *P. protegens* CHA0 that trigger induced systemic resistance in rice against *Cochliobolus miyabeanus* (Ma et al. 2017).

6.2.3. *Peptaibols*

Peptaibols are a large family of antimicrobial peptides of a short-chain length, typically between 8 and 20 residues that include an unusual amino acid (α-amino isobutyric acid, isovaline, ethylnorvaline or hydroxyproline), an acetylated N-terminus and a C-terminal amino alcohol. They are produced by fungi, including Trichoderma and related genera such as *Emericellopsis* and *Gliocladium*.

They are amphipathic membrane-active compounds with the ability to aggregate and form ion channels in lipid bilayer membranes (Marik et al. 2019). The genus Trichoderma produces alamethicin, peptaivirins, trichokonins or trichorzianines have been related to the biocontrol of plant diseases.

Few peptaibols have been assayed against plant pathogens. For example, trichokonin VI from *Trichoderma pseudokoningii* strain SMF2 induced resistance in tobacco and Chinese cabbage (Luo et al. 2010; Li et al. 2014b), and also induced extensive apoptotic programmed cell death in plant fungal pathogens (Shi et al. 2012). Other examples are trichokonins obtained from *Trichoderma koningii* (VI, VII and VIII) and trichorzianines isolated from *T. harzianum* (A1 and B1), which showed a broad antifungal activity (Khan et al. 2020), an 18-residue long peptaibol produced by *Trichoderma virens* Gv29-8 that seems to be critical in the activation of systemic protection to the leaf pathogenic bacteria in cucumber seedlings (Viterbo et al. 2007). In addition, peptaivirins (A and B) show antiviral activity and reduce tobacco mosaic virus infection in tobacco plants (Yun et al. 2000).

6.2.4. *Other types of peptides produced by microorganisms*

Pseudopeptides have few peptide bonds and complex amino acid modifications. There are several examples among actinomycetes with interest in plant disease control. The peptidyl nucleoside antibiotics are synthesized by *Streptomyces* spp. such as polyoxins and nikkomycins produced by *Streptomyces cacaoi* and *Streptomyces tendae* (Zhang et al. 2020). The polyoxins are effective in reducing the gray mold disease of grapes (*Botrytis cinerea*), the black spot disease of Japanese pear (*Alternaria kikuchiana*) and rice fungal diseases such as those caused by *Pyricularia oryzae* and *Cochliobolus miyabeanus* (Osada 2019). Polyoxin D is registered as a fungicide in USA and Canada.

Another example is the peptide antibiotics neopeptins isolated from *Streptomyces* sp. KNF2047 that showed disease control activity against powdery mildew on cucumber plants (Kim et al. 2007). Other interesting peptides from actinomycetes are the cyclic thiopeptide cyclothiazomycin B1 (CTB1 from *Streptomyces* sp. HA 125-40, cyclo (L-leucyl-L-prolyl) from *Streptomyces* sp. KH-614 (Rhee 2003), coronamycin from Streptomyces sp. MSU-2110, glomecidin from *Streptomyces lavendulae* H698 SY2 (Kunihiro et al. 2003) and kutznerides 1-4 from Kutzneria sp. 744 (Broberg et al. 2006).

Depsipeptides are a class of non-ribosomal peptides cyclized via an ester bond and often contain non-proteinogenous amino acids that are like CLPs. Pseudomonas spp. produce some depsipeptides such as nunamycin and nunapeptin that show potent activity against plant pathogens, in particular, the basidiomycete *Rhizoctonia solani* and *Pythium aphanidermatum* (Michelsen et al. 2015).

Another interesting peptidic antibiotic is pantocin A that is produced by strains of *Pantoea agglomerans* and involved in the biocontrol of *Erwinia amylovora*, the causal agent of fire blight of rosaceous plants (Smits et al. 2019).

Several fungi produce other types of peptides that are involved in biocontrol. For example, cyclic dipeptides or diketopiperazines such as chaetomin produced by *Chaetomium globosum* with antifungal activity against *Pythium* (Di Pietro et al. 1992), the cysteine-rich defensin-like peptides, antifungal protein (AFT) from *Aspergillus giganteus* with activity against *Magnaporthe grisea, Fusarium moniliforme and Phytophthora infestans* (Vila et al. 2001), PAF or PgAFP from *Penicillium chrysogenum*, ANAFP from *Aspergillus niger*, NAF from *Penicillium nalgiovense*, AcAFP and AcAMP from *Aspergillus clavatus* and NFAP from *Neosatorya fisheri* (Hegedüs and Marx 2012).

6.3. Peptides from plants

Plants have developed complex mechanisms to defend themselves from biotic and abiotic stresses by perceiving nonself molecular patterns associated with pathogens (PAMPs; e.g. bacterial flagella), or damage from injured host cells (DAMPs; e.g. breakdown products of plant extracellular matrix). This perception is through transmembrane pattern recognition receptors (PRRs, mainly receptor kinases and receptor-like proteins) that activate the plant immune system signaling cascade. The first line of inducible defense is pattern-triggered immunity (PTI) that involves common signaling molecules, for example, Ca^{2+}, ROS and hormones (primarily salicylic acid, SA, ethylene, ET, and jasmonic acid, JA), and extensive transcriptional reprogramming (Gust et al. 2017; Campos et al. 2018; Li et al. 2020, 2021a).

Pathogenic microbes produce polymorphic effector proteins that promote virulence through inhibition or mimic of plant functions. Plants counteract this susceptibility through resistance R proteins, which activate effector-triggered immune responses (ETI) (e.g. cell death, antimicrobial PR proteins, ROS, etc.) (Dangl et al. 2013).

Table 6.2 shows the main relevant AMPs and DAMPs.

Type	Family	Peptide	Amino acid number	Activity/plant species[x]
AMPs	Thionins	Purothionin, viscotoxin	45–48	Ab, Af
	Defensins	PhD1, Fa-AMP1.2, RsAFP1-4	45–54	Ab, Af, Pi
	Heveins	Hevein, Pn-AMP1.2h2, NTC	29–45	Af, AbG+, Aoo
	Knottins	PCI, Psacothensin, Mj-AMP1	28–37	Ab, Af, Pi, Av
	Hairpinins	Luffin P1, MBP-1	31–50	Ab, Af, Pi, Ri
	Lipid transfer proteins	wLTP2, Lc-LTP1-3, Ps-LTP1	70–90	Ab, Af
	Snakins	Snakin 1	60–70	Ab, Af, Am
	Cyclotides	Kalata, VavAF, PocaAB, Cy04	28–37	Ab, Av
DAMPs	Systemins	Systemin	18	So
	HP systemins	Hypr-systemin	15–20	So
	Plant elicitor peptides	Pp-PEP, Md-PEP, At-PEP	23–36	Ba, Po, Ro,So, Fab
	Inceptins	Inceptin	11	Vi, Fa
	Rapid alkalinization factors	RALFs	near 50	Widely
	Phytosulfokines PSKs	5	widely	Ab, Af, At
	PAMP induced peptides	PIP	11	At, widely
	CAP derived peptides	CAPE1	11	At, Sl
	Z. mays immune signaling peptide	Zip1	17	Zm

[x] Ab, antibacterial; Af, antifungal; AbG+, antibacterial against Gram positives; Aoo, anti-oomycetes; Av, antiviral; Pi, proteinase inhibitor; Ri, ribosomal inhibitor; At, Arabidopsis thaliana; Ba, Brassicaceae; Fa, Fabaceae; Po, Poaceae; Ro, Rosaceae; Sl, Solanum lycopersicum; So, Solanaceae; Vi, Vitaceae; Zm, Zea mays.

Compiled from: Huffaker et al. (2006), Bartels et al. (2013), Hou et al. (2014), Lacerda et al. (2014), Bartels and Boller (2015), Tam et al. (2015), Gust et al. (2017), Campos et al. (2018), Ruiz et al. (2018a), Foix et al. (2021) and Li et al. (2021).

Table 6.2. *Antimicrobial peptides (AMPs) that are active against plant pathogenic bacteria and fungi, and damage-associated molecular patterns (DAMPs) produced by plants*

6.3.1. *Antimicrobial peptides from plants (AMPs)*

Plant AMPs are characterized by 4–12 cysteines, which form disulfide bridges with a compact structure and a strong molecular stability and resistance to chemical and thermal denaturation. Plant AMPs are classified based on their disulfide bond patterns and structure in the families of thionins, defensins, heveins, knottins, cyclotides, lipid transfer proteins, snakins and hairpinins. Table 6.2 summarizes the main characteristics of each family; for further details, see reviews (Tam et al. 2015; Ageitos et al. 2017; Li et al. 2021).

AMPs from plants differ from animal AMPs, mostly in modes of action. Lipid transfer proteins bind fungal membrane lipids to inhibit pathogen penetration into the host cell, knottins inhibit proteases, heveins bind chitin in fungi and some defensins interact with eukaryotic receptor proteins to target intracellular functions. Most are inducible and can be tissue-specific, and to maximize resistance, complex mixtures of different AMPs frequently operate together.

Many plant-derived AMPs have strong activity against a wide range of plant pathogenic fungi and bacteria, and many transgenic plants expressing heterologous AMPs are protected from pathogens (Lacerda et al. 2014; Campos et al. 2018). Snakin-1 and snakin-2 inhibit the growth of fungi (e.g. *Magnaporthe grisea*, *Fusarium solani* and *Botrytis cinerea*) and bacteria (e.g. *Dickeya dadantii*, *Ralstonia solanacearum* and *Sinorhizobium meliloti*) (Berrocal-Lobo et al. 2002).

The exogenous application of Rs-AFP1 and Rs-AFP2 resulted in the protection of Arabidopsis plants, as well as the recombinant defensin Drr230a protected soybean against *Fusarium*, *Colletotrichum* and *Phakopsora* (Lacerda et al. 2016), and the peptide MaSAMPs from *Microcitrus australasica* protected citrus against Huanglongbing disease (Huang et al. 2021) and thionines confer resistance to Citrus canker (Hao et al. 2016).

The expression of plant AMPs is also active against bacterial and fungal plant pathogens (Ali et al. 2018; Sher Khan et al. 2019). For example, in cereals, the expression of radish Rs-AFP2 confers resistance against *Magnaporthe oryzae* and *Rhizoctonia solani* in rice (Jha and Chattoo 2010), of LTP against *Cochiolobus sativus* and *Fusarium graminearum* on wheat (Zhu et al. 2012) and of snakins from potato in front of *Blumeria graminis* on wheat (Facció et al. 2022). In tobacco, the expression of defensins protected against sheath blight disease (Wang et al. 2017), hevein from Pharbitis nilPn-AMP protected against *Phytophthora parasitica* (Koo et al. 2002) or the soybean γ-thionin-SE60 resulted in enhanced resistance against *Pseudomonas syringae* (Choi et al. 2008). In bananas, the expression of petunia defensins increased the resistance against *F. oxysporum* (Ghag et al. 2012).

In Solanaceae, the expression of snakines was effective against *Clavibacter michiganensis* or *Pectobacterium atrosepticum* in tomato and potato (Mohan et al. 2014), that of thionins from Brassicaceae in potato against *Botrytis cinerea* and overexpression of *Mirabilis jalapa* defensin Mj-AMP1 in tomato enhanced the resistance to *Alternaria solani* (Schaefer et al. 2005).

6.3.2. *Damage-associated molecular patterns (DAMPs)*

They are short peptides cleaved out of precursor proteins, usually upon damage perception. Table 6.2 summarizes some characteristics of the most relevant families. For further details, see recent interesting reviews, such as Gust et al. (2017), Hou et al. (2019) and Li et al. (2020) and references therein. DAMPs are ubiquitously distributed in plants and secreted to the apoplast, which occurs through the conventional ER pathway. Apoplastic mature secondary DAMPs are perceived by specific PRRs, in general LRR-RLK, and upon binding, they further interact with LRR-RLK coreceptors (e.g. Bak1) to initiate downstream signaling. Pep-PEPR complexes and the plant response to Peps have been thoroughly analyzed (Bartels 2013, 2015) in *Arabidopsis* (Ross 2014) and *Prunus* (Foix 2021). The responses include phosphorylation events, Ca2+-dependent signaling, production of nitric oxide (NO) and ROS, activation of MAPKs, co-activation of SA, JA and ET hormone pathways (Huffaker 2011) or production of secondary metabolites. PROPEPs and Peps have been identified in most angiosperm plant species like Brassicaceae, Poaceae, Rosaceae, Poaceae, Fabaceae and Solanaceae. Each plant family is characterized by a specific Pep consensus motif, and Pep and PEPR sequences from the same family are compatible (Lori 2015; Ruiz et al. 2018b).

Topical application of synthetic peach PpPep1 and PpPep2 reduced the severity of *Xanthomonas vesicatoria* pv. *arboricola* infection in peach plants (Ruiz et al. 2017). In *Arabidopsis*, AtPep pretreatment or overexpression of AtPROPEP1 or AtPROPEP2 increases resistance to *Pst* DC3000 and *Pythium irregulare* (Huffaker et al. 2006). Externally applied PIP1 and PIP2 lead to enhanced immune responses and resistance to *Pseudomonas syringae* and *Fusarium oxysporum* (Hou et al. 2014).

The most interesting aspect of DAMPs is that they are active at nanomolar concentrations, compared to AMPs needed at micromolar concentrations, thus, emphasizing their interest in plant protection.

6.4. Peptides from animals

Peptides in animals have an essential role in defense, on the one hand, protecting directly against pathogens due to their antimicrobial activity, and on the other hand, by modulating the immunological response, especially in vertebrates. In vertebrates, the AMPs are the first innate protection barrier of the defense system (Hancok and Scott 2000; Brogden 2003; Mansour et al. 2014; Kumar et al. 2020). In addition to the defense role, the peptides play a role in the symbiotic interactions between microorganisms and animals (Maróti et al. 2011).

6.4.1. *Antimicrobial peptides in animals*

From the vast amount of AMPs produced by the animals, the most relevant examples are in poriphers, cnidarians and molluscs like discodermin A, defensins, mytilins and myticins; in arthropods, cecropin, mellitin, thanatin and apidaecins; in fishes, cathelidicins, beta defensins, hepcidins and piscidins; in amphibians, bombinins, maximins, cathelidicins, esculentins, dermaseptins, phylloseptins, palustrin, magainin, pseudins and ranaturerins; in birds and reptiles, they are important in cathelidicins (OH-CATH, Fowlicidins, CATH) and beta defensins (TBD-1, crotamine, TEWP); and finally in mammals, the cathelidicins (Ll-37, indolicidin, protegrins, bactenecins), defensins (α (HPN-1, RatNP1), β (AP, LAP, RTD, BNBD) and θ (rtd-1 TO 3), dermicidin, hepcidin and hepticin (Ageitos et al. 2016).

Some of the most relevant AMPs of animal origin are the cecropins, melittins, magainins, defensins and cathelicidins. Cecropins are AMPs from insects that interact with pathogen cell membranes causing lysis first identified in the hemolymph of the giant silk moth. Cec A is α-helical structure mainly active against Gram-negative bacteria (Lee and Lee 2015). Cecropin P1 has also been reported in the nematode *Ascaris suum*, a parasite of the pig intestine (Brogden et al. 2003). Melittin is another AMP from bee venom with an α-helical structure and has a strong activity against bacteria, fungi and virus, but it is highly hemolytic (Wu et al. 2018). The magainins originate from amphibians; the most studied are from the African clawed frog skin and cause cell lysis, and also interfere in DNA synthesis and metabolic processes in bacteria and some fungi (McMillan and Coombs 2020). In the case of β-defensins and catheledicins, in addition to the lytic effect, they have intracellular targets by inhibiting replication, translation and ion channels (Kościuckuz et al. 2012; Kumar et al. 2020).

6.4.2. Peptides from animal origin active against plant pathogens and diseases

Several AMPs from animal origin have been tested against plant pathogenic bacteria and fungi or to control plant diseases caused by them. Most studies deal with activity in vitro, and less were done in interaction with plants, while several have been done with analogs of AMPs of animal origin (see below).

There are several studies, involving peptides from crustaceans (tachyplesin I), sheep (SMAP-29), humans (histatin-5, LL37), cows (indolicidin), insects (cecropins A and B, apidaecin IA, drosocin, pyrrhocoricin) and frogs (magainin and melittin) tested against *Agrobacterium tumefaciens* or *Xanthomonas citri* ssp. citrí (Stover et al. 2013). Other examples include lactoferricin from cows against *Pseudomonas syringae*, *Xanthomonas campestris* and *Ralstonia solanacearum* (Fukuta et al. 2012).

Plant assays are more limited, and have used different strategies either by topical application or by heterologous expression of AMP genes in the plant. Topical application by spraying the AMP has been studied with gomesin against *Xylella fastidiosa* in tobacco (Arora et al. 2018), and Hylin-a1 and tripticin against the Citrus canker disease in citrus (Kisi et al. 2018).

However, the most frequent strategy used with peptides from animals has been the heterologous expression in transgenic plants for resistance to fungal and bacterial diseases, using plant species of important crops. Table 6.3 shows several recent examples like in Chinese cabbage expressing human cathelicidin LL-37 that was protected against *Fusarium oxysporum* f.sp. lycopersici, *Colletotrichum higginsianum* and *Rhizoctonia solani* (Jung et al. 2012); tomato plants expressing cecropin B resistant to bacterial wilt and spot; pearl millet expressing magainin-2 from frogs against downy mildew (Ramadevi et al. 2014); barley expressing metchnikowin from fly front powdery mildew (Rahnamaelan and Vilcinskas 2012); grass expressing Penaeidin 4-1 from fly confer resistance to dollar spot and brown patch (Zhou et al. 2011); tobacco expressing cow lactoferricin more resistant to *P. syringae* pv. tabacci and *Botrytis cinerea* (Fukuta et al. 2012); and rice and wheat expressing thanatin resistant to rice blast and aflatoxins producer fungus in wheat (Shubert et al. 2015). Other examples are in citrus and fruit trees as the expression of attacin A from cabbage looper, dermaseptin from frogs and sarcotoxin IA from flies in sweet orange against the Citrus canker disease (Cardoso et al. 2010; Furman et al. 2013; Kovayashi et al. 2017).

Antimicrobial peptide	Amino acid number	Sequence	Source
Brevinin-2SSb	40	SLFSLIKAGAKFLGKNMLKQGPQYPACKVSKDSENVNWKS	*Glandirana susurra*
Cathelicidin LL-37	37	MLLGDFFRKSKEKIGKEFKRIVQRIKDFLRNLVPRTES	*Homo sapiens*
Cecropin B	35	KWKVFKKIEKMGRNIRNGIVKAGPAIAVLGEAKAL	*Hyalophora cecropia*
Dermaseptin	35	MALWKTMLKKLGTMALHAGKAALGAAADTISQGTQ	*Phyllomedusa sauvagii*
Gomesin	18	ZCRRLCYKQRCVTYCRGR	*Acanthoscurria gomesina*
Granuliberin-Ssa	12	FIFLPIFRRPVS	*Glandirana susurra*
Hylin-a1	17	IFGAILPLALGALKNLIK	*Hypsiboas albopunctatus*
Lactoferricin	25	FKCRRWQWRMKKLGAPSITCVRRAF	*Cow's milk*
Magainin-2	23	GIGKFLHSAKKFGKAFVGEIMNS	*Xenopus laevis*
Mellitin	26	GIGAVLKVLTTGLPALISWIKRKRQQ	*Apis mellifera*

Antimicrobial peptide	Amino acid number	Sequence	Source
Metchnikowin	26	HRHQGPIFDTRPSPFNPNQPRPGPIY	*Drosophila melanogaster*
Ranatuerin-2SSa	28	GLISTIWNTASNVAGTLTDSVKCKFKKC	*Glandirana susurra*
Penaeidin4-1	47	HSSGYTRPLRKPSRPIFIRPIGCDVCYGIPSSTARLCCFRYGDCCHL	*Litopenaeus setiferus*
Sarcotoxin IA	63	MNFQNIFIFV ALILAVFAGQ SQAGWLKKIG KKIERVGQHT RDATIQGLGI AQQAANVAAT ARG	*Sarcophaga peregrina*
Scorpin like	96	MNAKITVLIFLVAITIASCGWISEKKAQDFVDSKLPNGMVKNAIKAVIHKASK NKHGCFADIDVGGGCTSHCKKEDKEGICHGTKCKCGIPRAYKK	*Vaejovis mexicanus*
Thanatin	21	GSKKPVPIIYCNRRTGKCQRM	*Podisus maculiventris*
Tritrpticin	13	VRRFPWWWPFLRR	*Sus domesticus*

Compiled from: Hancock and Scott (2000), Brogden et al. (2003), Maróti et al. (2011), Zhou et al. (2011), Fukuta et al. (2012), Kościuckuz et al. (2012), Furman et al. (2013), Mansour et al. (2014), Lee and Lee (2015), Kobayashi et al. (2017), Kumar et al. (2020) and McMillan and Coombs (2020).

Table 6.3. *Peptides of animal origin with activity against plant pathogenic fungi or bacteria*

There are other interesting approaches of the use of heterologous expression of AMPs from animal origin, with agronomical projection, such as the expression of cecropin B in the scion of grapevines that protects the grafted variety from Pierce's disease; it is the so-called graft-mediated resistance (Dandekar et al. 2019). Another innovative strategy is the so-called paratransgenic strategy, applied to plant diseases transmitted by insect vectors, as is the case of the control of *Xylella fastidiosa*, by means of the transformation of the endogenous bacterium *Pantoea agglomerans* (an inhabitant of the gut insect) with a vector expressing scorpion and magainin peptide genes (Arora et al. 2018).

6.5. Synthetic peptides

Although peptides from living organisms are of natural origin, most are not exempt from toxicity or negative environmental impact or are found in very low concentrations. Therefore, in many cases, they are required to be synthesized or modified to improve several properties like decreasing toxicity, increasing stability, improving their activity or finding the minimal domain still maintaining their activity.

6.5.1. *Development of synthetic peptides*

The process of development of synthetic peptides consists of a first stage of generating the peptide sequences using a rational design, generally departing from natural sequences, or preparing newly designed compounds. To build up the peptide library, the synthesis is performed by solid phase methods (e.g. O-rink FMOC). Frequent approaches consist of increasing the amphiphilicity of an alpha-helical peptide, repeat sequences in tandem, cyclization or addition of particular end sequences or amino acids (e.g. D-amino acids). In any case, the objective of the first stage is to optimize their properties (antibacterial, antifungal, lytic, toxicity, resistance to proteases, stability in adverse environments) and the generation of libraries that can be entered into a screening process (Marcos et al. 2008; Rajaseharan et al. 2013).

An example of the template-based design using as lead natural peptides or consensus sequences from a family of peptides is the CECMELL11 library of 125 members of linear undecapeptides based on cecropin A and melittin fragments, with the bacteriolytic and cell-penetrating peptide BP100

(KKLFKKILKYL) (Badosa et al. 2007). In other cases, the modification consisted of a chemical modification of existing compounds (halogenation, cyclization, capping, conjugation, peptoids), or the use of particular motifs at the end, in chain or repeated sequences (ATCUN, Rana box, LPS binding gamma-core motifs) (Andreev et al. 2018). An example of newly designed was the cyclic decapeptide library CYCLO10 from which the peptide BPC194 (c(KKLKKFKKLQ)) had a potent antibacterial activity with low toxicity (Monroc et al. 2006; Cirac et al. 2017).

6.5.2. *Synthetic peptides against plant pathogens and diseases*

There are a vast number of reports on synthetic peptides active against plant pathogenic bacteria and fungi, and we only refer here to the most recent ones. For a more detailed description, the reader is referred to the reviews (Marcos et al. 2008; Montesinos 2012).

However, most of the studies deal with in vitro activity, less within planta studies and field studies are very scarce.

Table 6.4 includes a summary of the synthetic peptides that have been found active against plant pathogenic bacteria and fungi. Most abundant analogs are derived from animal peptides like cecropins and dermaseptins, followed by indolicidin, flagellin, histatin, hyalin, lactoferricin, LK peptide, magainin, ocellatin, scorpion and thanatin, as well as de novo designed.

Among the cecropin analogs, the peptides D5C and D2A21 are effective against oomycetes. Several hybrid peptides of short cecropin sequences and of other peptides (e.g. melittin, magainin) have also been developed. For example, a CECMEL11 library of linear undecapeptides made of combinatorial chemistry from cecropin and melittin fragments provided 125 members, some of them with a huge range of activity.

Interestingly, BP15 has antifungal activity against several post-harvest fungi; BP100 is antibacterial and active against the plant pathogenic bacteria *E. amylovora*, *Xanthomonas* spp. and *Pseudomonas syringae* pvs., and BP178 (a BP100 derivative with a magainin sequence and adapted to plant expression) shows antibacterial activity against *Xylella fastidiosa* (Baró et al. 2020a, 2020b) and plant defense elicitation activities (Montesinos 2021).

Original peptide	Analog	Amino acid number	Sequence	Activity against[x]
Cecropins	D5C	27	KRKRAVKRVGRRLKKLARKIARLGVAF	Aoo
	D2A21	23	FAKKFAKKFKKFAKKFAKFAFAF	Ab, Af
	BP100	11	KKLFKKILKYL-NH_2	Ab
	BP15	11	KKLFKKILKVL-NH_2	Af
	CA-M	15	KWKLFKKIGAVLKVL-NH_2	Ab
	MrsA1	34	MALEHMKWKLFKKIGIGAVLKVLTTGLPALKLTK	Af
	BP178	29	KKLFKKILKYLAGPAGIGKFLHSAKKDEL	Ab/Pe
	RW-BP100	11	RRLFRRILRWL-NH_2	Ab/Af
	KSL-W	11	KKVVFWVKFK-NH_2	Ab/Af
Dermaseptins	MsrA2	32	MAMWKDVLKKIGTVALHAGKAALGAVADTISQ	Ab/Af
	CBD-DrsB1	34	ALWKTMLKKLGTMALHAGKAALGAAADTISQGTQ	Af
Indolicidin	10R	13	RRPWKWPWWPWRR	Af
Flagellin	Flg15	15	RINSAKDDAAGLQIA-NH2	Pe
Histatin	Dhvar-5	14	LLLFLLKKRKKRKY-NH_2	Ab
Hyalin-a1	K0-W6-Hy-a1	19	KIFGAIWPLALGALKNLIK	Ab

Original peptide	Analog	Amino acid number	Sequence	Activity against[x]
Lactoferricin	JH8944	13	FKCKKVVISLRRY	Af
LK peptide	3.1	11	KKLLKWLLKLL-NH_2	Ab
Magainin	SP13-D	20	KR*R*LIA*R*I*L*RLAARA*L*V*K*KR	Af
Ocellatin	Ocellatin-4	21	KLLKFVTKVGKAIFKALIKAI	Af
Scorpion IsCT	SP1-1, SP10-D	12	RKKRLKLLKRLL	Ab/Af
Thanatin	ThanatinS	21	GSKKPVPIIYCNRRSGKCQRM	Ab/Af
De novo designed	CYCLO10 (BPC194)	10	c(KKLKKFKKLQ)	Ab/Af
	D4E1	18	FKLRAKIKVRLRAKIKL-NH_2	Aoo
	PAF26	6	RKKWFW	Af
	RIJK2	12	RIVWVRIRRWFV-NH_2	Ab/Abi

[x] Ab, antibacterial; Af, antifungal; Aoo, anti-oomycetes; Pe, plant defense elicitor; Abi, antibiofilm.

Compiled from: Marcos et al. (2008), Jha and Chattoo (2010), Güell et al. (2011), Badosa et al. (2013), Ng-Choi et al. (2014), Lee and Kim (2015), Puig et al. (2015), Badosa et al. (2017), Camó et al. (2017), Cirac et al. (2017), Hao et al. (2017), Rufo et al. (2017), Kishi et al. (2018), Oliveras et al. (2018), Dandekar et al. (2019), Baró et al. (2020a), Baró et al. (2020b), Kundu (2020), Caravaca-Fuentes et al. (2021), Huang et al. (2021), Moll et al. (2021) and Montesinos et al. (2021).

Table 6.4. *Synthetic peptides and analogs active against plant pathogenic bacteria and fungi*

Others such as 10R (indolicidin analog) and JH8944 (lactoferricin analog) show antifungal activity, SP1 and Sp10 (scorpion IsCT analogs) are antibacterial and antifungal, whereas 3.1 (LK peptide analog) and Dhvar-5 (histatin analog) and the synthetic cyclodecapeptide BPC194 from the CYCLO10 library show antibacterial activity (Monroc et al. 2006).

Interestingly, a group of synthetic peptides have shown antibiofilm activity, and a group was particularly active against *Xylella fastidiosa* (Moll et al. 2021).

Topical application of synthetic peptides like BP100 have been reported to be effective against infection by stolbur phytoplasm in periwinckle (Rufo et al. 2017), and several plant pathogenic bacteria such as *Erwinia amylovora* in pear, *Xanthomonas vesicatoria* in pepper and *Pseudomonas syringae pv.* tomato in tomato (Badosa et al. 2017). Others such as BP15 have been reported to be effective in preventing fungal rot infections by *Penicillium expansum* in apples (Badosa et al. 2009). However, the most relevant study was reported with peptide BP15 in a field study with significant control levels of brown spot disease of pear (caused by the fungus *Stemphylium vesicarium*) (Puig et al. 2015).

Studies involving transgenic plants expressing synthetic peptides are also abundant, but since there are analogs of animal, microbial or plant peptides, they are described in the corresponding section.

6.6. Biotechnological production of peptides

Heterologous expression in living systems as biofactories appears more sustainable than chemical synthesis, providing a promising tool for the production of functional peptides. However, this approach is only achievable with linear AMPs composed of proteinogenous amino acids (Parachin et al. 2012). In addition, it often requires the re-design of the peptide due to the need for including sequences to adapt the resulting gene construct to the requirements of transcription or translation in a particular living host (e.g. addition of sequences for expression, processing, intracellular trafficking, etc.), and this approach can lead to a loss of activity of the resulting peptide (Nadal et al. 2012; Badosa et al. 2013).

The design of an expression strategy to successfully produce a functional peptide requires solving problems related to the small size in certain peptides, low yield, degradation due to intracellular proteases, toxicity towards host cells (e.g. an antimicrobial in a microbial platform) and the high production costs associated with the extraction and purification. Many peptides and proteins have been expressed in microbial systems like *E. coli* and *Pichia pastoris* and plants. More complex systems such as mammalian or insect cell lines are not or less used due to the high production costs, time-consuming process and difficulties in scaling up the recombinant protein recovery.

6.6.1. *Microbial systems*

Most recombinant peptides have been produced using a microbial platform as a biofactory, being *Escherichia coli, Pichia pastoris* and *Saccharomyces cerevisiae*, the most commonly used expression system (Deng et al. 2017).

Among microbial platforms, bacterial systems are frequently used since they are simple to manipulate, often provide high peptide yield and are easy to handle. However, they present serious disadvantages related to the inclusion body formation (protein bodies) when produced in the cytoplasm, peptide degradation, inability to perform the post-translational modifications sometimes and the need to remove endotoxins produced by the microorganism. The classical approach based on the expression of heterologous proteins in the cytosol has been replaced by the polypeptide expression and secretion into the extracellular media that in the case of peptides greatly simplifies the downstream recovery process. The yeast system *Pichia pastoris* is a suitable system for high density cultivation, allowing strong secretion due to its eukaryotic machinery, reducing the downstream purification processing costs because it secretes only a few host cell proteins into the fermentation broth (Kuddus 2016; Kastilan et al. 2017). This system is advantageous compared to *Saccharomyces cerevisiae* because of the availability of strong inducible promoters. For example, ch-penaedin (CHP) was successfully produced with a high recombinant AMP yield of about 108 mg/L (Li et al. 2005). Expression of several AMPs in *E. coli* failed, while it was successfully produced in *P. pastoris* (Deng et al. 2017). However, advantages of AMP expression in *E.coli* over the other microbial systems have also been described. For example, NUMAtag uses a secretion strategy based on *E. coli* hemolysin A (HlyA) type 1 secretion system (T1SS) to produce heterologous peptides and proteins with elevated secretion levels (in the range of mg/L), and without residual amino acids of the signal sequence in the final product (Khosa et al. 2017).

6.6.2. *Microalgae systems*

Several recombinant proteins are expressed in algal chloroplasts, although in general. Low expression levels are obtained in comparison to systems like chloroplast of higher plants. The unicellular algae *Chlamydomonas reinhardtii* emerged as a model of production platform that has a short doubling time, and its three genomes are sequenced (chloroplast, nuclear and mitochondrial). Chloroplasts become the most interesting organelle to produce recombinant proteins due to the high capacity to accumulate proteins and a high copy number of the chloroplast genome, but much effort is necessary to make the algal chloroplast platform commercially competitive. In the case of peptides, a mutated defensin, mNP-1, was produced in *Chlorella ellipsoidea* at a yield of 100 mg/g (Fan et al. 2019) and cAFLPm3 in *Chlamydomonas reinhardtii* at a yield of 0.35% of total soluble protein (Li et al. 2021).

6.6.3. *Plant systems*

Transgenic plants offer several advantages for the large-scale production of high-value recombinant proteins and AMPs (Sharma and Sharma 2009). Plant biofactories are potentially one of the most economical systems of low production costs, easy scaling up or down, and with relatively simple purification technology (Shanmugaraj et al. 2021). This system offers multiple possibilities for the expression of AMP genes as different tissues can be chosen, several promoters and strategies to reduce transgene product degradation and toxicity against plants cells, while increasing its stability to obtain high yield production. Objections such as long times to product obtainment, regulatory framework, post-translational processing or low expression levels prevent their commercialization, but major advances have been conducted to overcome these limitations (Guan et al. 2013).

Plant systems for the production of AMPs include cereal seeds, leaf chloroplasts, cells, root tissues or whole plants, among others, and different production yields were reported depending on the platforms used, the strategy and the physicochemical properties of the AMP. In cereal seeds as biofactories, the accumulation of chimeric rh-LL37 at 0.55 μg/g of seed and

MBP-rhLL-37 at 250 mg/g of spike tissue in barley (Holásková et al. 2018; Mirzaee et al. 2021), BP178, CecA and PAF102 at 20–38 µg/g of rice seeds were reported (Montesinos et al. 2016, 2017; Bundó et al. 2019). Using tandem multimeric expression (29 multimers), a yield of 2 mg/g of seed of lactostatin peptide was reported in rice seeds (Cabanos et al. 2012). In the *Nicotiana benthamiana*-platform, stable transformation of leaves yielded 6–7 µg/g f.w. of the peptide MsrA2 and 5.5–6 µg/g f.w. of DrsB1, while transient expression enables the production of AfpB and SP1-1 at a yield of 225 and 25 µg/g f.w., respectively (Zeitler et al. 2013; Shi et al. 2019).

Therefore, plants not only appear as a viable alternative system to produce AMPs, but also as a cost-effective delivery platform, conferring protection against plant pathogens.

6.7. References

Abriouel, H., Franz, C.M., Ben Omar, N., Gálvez, A. (2011). Diversity and applications of *Bacillus bacteriocins*. *FEMS Microbiol. Rev.*, 35, 201–232.

Ageitos, J., Sanchez-Perez, A., Calo Mata, P., Villa, T.G. (2017). Antimicrobial peptides (AMPs): Ancient compounds that represent novel weapons in the fight against bacteria. *Biochem. Pharmacol.*, 133, 117–138.

Ali, S., Ganai, B.A., Kamili, A.N., Bhat, A.A., Mir, Z.A., Bhat, J.A., Tyagi, A., Islam, S.T., Mushtaq, M., Yadav, P. et al. (2018). Pathogenesis-related proteins and peptides as promising tools for engineering plants with multiple stress tolerance. *Microbiol. Res.*, 29–37, 212–213.

Arora, A.K., Pesko, K.N., Quintero-Hernández, V., Possani, L.D., Miller, T.A., Durvasula, R.V. (2018). A paratransgenic strategy to block transmission of *Xylella fastidiosa* from the glassy-winged sharpshooter *Homalodisca vitripennis*. *BMC Biotechnol.*, 18, 1–10.

Badosa, E., Ferré, R., Planas, M., Feliu, L., Besalú, E., Cabrefiga, J., Bardají, E., Montesinos, E. (2007). A library of linear undecapeptides with bactericidal activity against phytopathogenic bacteria. *Peptides*, 28, 2276–2285.

Badosa, E., Ferré, R., Francés, J., Bardají, E., Feliu, L., Planas, M., Montesinos, E. (2009). Sporicidal activity of synthetic antifungal undecapeptides and control of Penicillium-rot of apples. *Appl. Environ. Microbiol.*, 75, 5563–5569.

Badosa, E., Moiset, G., Montesinos, L., Talleda, M., Bardají, E., Feliu, L., Planas, M., Montesinos, E. (2013). Derivatives of the antimicrobial peptide BP100 for expression in plant systems. *PloS One*, 8, e85515.

Badosa, E., Montesinos, L., Camó, C., Ruz, L., Cabrefiga, J., Francés, J., Gascón, B., Planas, M., Feliu, L., Montesinos, E. (2017). Control of fire blight infections with synthetic peptides that elicit plant defence responses. *J. Plant Pathol.*, 99, 65–73.

Baró, A., Badosa, E., Montesinos, L., Feliu, L., Planas, M., Montesinos, E., Bonaterra, A. (2020a). Screening and identification of BP100 peptide conjugates active against *Xylella fastidiosa* using a viability-qPCR method. *BMC Microbiol.*, 20, 1–14.

Baró, A., Mora, I., Montesinos, L., Montesinos, E. (2020b). Differential susceptibility of *Xylella fastidiosa* strains to synthetic bactericidal peptides. *Phytopathology*, 110, 1018–1026.

Bartels, S. and Boller, S. (2015). Quo vadis, Pep? Plant elicitor peptides at the crossroads of immunity, stress, and development. *J. Exp. Bot.*, 66, 5183–5193.

Bartels, S., Lori, M., Mbengue, M., van Verk, M., Klauser, D., Hander, T., Boni, R., Robatzek, S., Boller, T. (2013). The family of peps and their precursors in *Arabidopsis*: Differential expression and localization but similar induction of pattern-triggered immune responses. *J. Exp. Bot.*, 64, 5309–5321.

Beric, T., Kojic, M., Stankovic, S., Topisirovic, L., Degrassi, G., Myers, M., Venturi, V., Fira, D. (20212). Antimicrobial activity of *Bacillus* sp. natural isolates and their potential use in the biocontrol of phytopathogenic bacteria. *Food Technol. Biotechnol.*, 50, 25–31.

Berrocal-Lobo, M., Segura, A., Moreno, M., López, G., García-Olmedo, F., Molina, A. (2002). Snakin-2, an antimicrobial peptide from potato whose gene is locally induced by wounding and responds to pathogen infection. *Plant Physiol.*, 128, 951–961.

Broberg, A., Menkis, A., Vasiliauskas, R. (2006). Kutznerides 1-4, depsipeptides from the actinomycete *Kutzneria* sp. 744 inhabiting mycorrhizal roots of *Picea abies* seedlings. *J. Nat. Prod.*, 69, 97–102.

Brogden, K.A., Ackermann, M., McCray, P.B., Tack, B.F. (2003). Antimicrobial peptides in animals and their role in host defences. *Int. J. Antimicrob. Agents*, 22, 465–478.

Bundó, M., Montesinos, L., Izquierdo, E., Campo, S., Mieulet, D., Guiderdoni, E., Rossignol, M., Badosa, E., Montesinos, E., San Segundo, B. et al. (2014). Production of cecropin A antimicrobial peptide in rice seed endosperm. *BMC Plant Biol.*, 14, 102.

Bundó, M., Shi, X., Vernet, M., Marcos, J.F., López-García, B., Coca M. (2019). Rice seeds as biofactories of rationally designed and cell-penetrating antifungal PAF peptides. *Front. Plant Sci.*, 10, 731.

Cabanos, C., Ekyo, A., Amari, Y., Kato, N., Kuroda, M., Nagaoka, S., Takaiwa, F., Utsumi, S., Maruyama, N. (2013). High-level production of lactostatin, a hypocholesterolemic peptide, in transgenic rice using soybean A1aB1b as carrier. *Transgenic Res.*, 22, 621–629.

Camo, C., Torné, M., Besalú, E., Rosés, C., Cirac, A.D., Moiset, G., Badosa, E., Bardaji, E., Montesinos, E., Planas, M. et al. (2017). Tryptophan-containing cyclic decapeptides with activity against plant pathogenic bacteria. *Molecules*, 22, 1–13.

Campos, M., de Souza, C., Sampaio de Oliveira, K., Campos Dias, S., Franco, O. (2018). The role of antimicrobial peptides in plant immunity. *J. Exp. Bot.*, 69, 4997–5011.

Cao, H., Ke, T., Liu, R., Yu, J., Dong, C., Cheng, M., Huang, J., Liu, S. (2015). Identification of a novel proline-rich antimicrobial peptide from *Brassica napus*. *PLoS One*, 10, 1–13.

Caravaca-Fuentes, P., Camó, C., Oliveras, À., Baró, A., Francés, J., Badosa, E., Planas, M., Feliu, L., Montesinos, E., Bonaterra, A. (2021). A bifunctional peptide conjugate that controls infections of *Erwinia amylovora* in pear plants. *Molecules*, 26, 1–14.

Cardoso, S.C., Barbosa-Mendes, J.M., Boscariol-Camargo, R.L., Christiano, R.S.C., Filho, A.B., Vieira, M.L.C., Mendes, B.M.J., Filho, F.D.A.A.M. (2010). Transgenic sweet orange (*Citrus sinensis* L. Osbeck) expressing the attacin a gene for resistance to *Xanthomonas citri* subsp. *citri*. *Plant Mol. Biol. Rep.*, 28, 185–192.

Caulier, S., Nannan, C., Gillis, A., Licciardi, F., Bragard, C., Mahillon, J. (2019). Overview of the antimicrobial compounds produced by members of the *Bacillus subtilis* group. *Front. Microbiol.*, 10, 302.

Chen, X., Hu, Y.H., Chen, W.D., Li, W.D., Huang, Z.C., Li, Y., Luo, Y.W., Huang, Y.X., Chen, Y.T., Wang, K. et al. (2016). Comparison of inducible versus constitutive expression of plectasin on yields and antimicrobial activities in *Pichia pastoris*. *Protein Expr Purif.*, 118, 70–76.

Choi, Y., Choi, Y.D., Lee, J.S. (2008). Antimicrobial activity of γ-thionin-like soybean SE60 in *E. coli* and tobacco plants. *Biochem. Biophys. Res. Commun.*, 375, 230–234.

Cirac, A.D., Torné, M., Badosa, E., Montesinos, E., Salvador, P., Feliu, L., Planas, M. (2017). Rational design of cyclic antimicrobial peptides based on BPC194 and BPC198. *Molecules*, 22, 7.

Company, N., Nadal, A., La Paz, J.L., Martínez, S., Rasche, S., Schillberg, S., Montesinos, E., Pla, M. (2014). The production of recombinant cationic α-helical antimicrobial peptides in plant cells induces the formation of protein bodies derived from the endoplasmic reticulum. *Plant Biotechnol. J.*, 12, 81–92.

Dandekar, A.M., Gouran, H., Ibanez, A.M., Uratsu, S.L., Aguero, C.B., McFarland, S., Borhani, Y., Feldstein, P.A., Bruening, G., Nascimento, R. et al. (2012). An engineered innate immune defense protects grapevines from Pierce disease. *Proc. Natl. Acad. Sci. U.S.A.*, 109, 3721–3725.

Dangl, J., Horvath, D., Staskawicz, B. (2013). Pivoting the plant immune system from dissection to deployment. *Science*, 341, 746–751.

Daniel, J.F. and Filho, E.R. (2007). Peptaibols of *Trichoderma*. *Nat. Prod. Rep.*, 24, 1128–1141.

Deng, T., Ge, H., He, H., Liu, Y., Zhai, C., Feng, L., Yi, L. (2017). The heterologous expression strategies of antimicrobial peptides in microbial systems. *Protein Expr. Purif.*, 140, 52–59.

Di Pietro, A., Gut-Rella, M., Pachlatko, J.P., Schwinn, F.J. (1992). Role of antibiotics produced by *Chaetomium globosum* in biocontrol of *Pythium ultimum*, a causal agent of damping-off. *Phytopathology*, 82, 131–135.

Eggenberger, K., Mink, C., Wadhwani, P., Ulrich, A.S., Nick, P. (2011). Using the peptide BP100 as a cell-penetrating tool for the chemical engineering of actin filaments within living plant cells. *Chembiochem.*, 12, 132–137.

Faccio, P., Vazquez-Rovere, C., Hopp, E., González, G., Décima-Oneto, C., Favret, E., Diaz Paleo, A., Franzone, P. (2013). Increased tolerance to wheat powdery mildew by heterologous constitutive expression of the *Solanum chacoense* snakin-1 gene. *Czech J. Genet. Plant Breed.*, 47, 135–141.

Fan, Y., Du, K., Gao, Y., Kong, Y., Chu, C., Sokolov, V., Wang, Y. (2013). Transformation of LTP gene into *Brassica napus* to enhance its resistance to *Sclerotinia sclerotiorum*. *Russ. J. Genet.*, 49, 380–387.

Flury, P., Vesga, P., Péchy-Tarr, M., Aellen, N., Dennert, F., Hofer, N., Kupferschmied, K.P., Kupferschmied, P., Metla, Z., Ma, Z. et al. (2017). Antimicrobial, insecticidal: Cyclic lipopeptides and hydrogen cyanide produced by plant-beneficial *Pseudomonas* strains CHA0, CMR12a, and PCL1391 contribute to insect killing. *Front. Microbiol.*, 8, 100.

Fogliano, V., Ballio, A., Gallo, M., Woo, S., Scala, F., Lorito, M. (2002). *Pseudomonas* lipodepsipeptides and fungal cell wall-degrading enzymes act synergistically in biological control. *Mol. Plant Microbe Interact.*, 15, 323–333.

Foix, L., Nadal, A., Zagorščak, M., Ramšak, Ž., Esteve-Codina, A., Gruden, K., Pla, M. (2021). *Prunus persica* plant endogenous peptides PpPep1 and PpPep2 cause PTI-like transcriptome reprogramming in peach and enhance resistance to *Xanthomonas arboricola* pv. *pruni*. *BMC Genomics*, 22, 360.

Fukuta, S., Kawamoto, K.-I., Mizukami, Y., Yoshimura, Y., Ueda, J.-I., Kanbe, M. (2012). Transgenic tobacco plants expressing antimicrobial peptide bovine lactoferricin show enhanced resistance to phytopathogens. *Plant Biotechnol.*, 29, 383–389.

Furman, N., Kobayashi, K., Zanek, M.C., Calcagno, J., García, M.L., Mentaberry, A. (2013). Transgenic sweet orange plants expressing a dermaseptin coding sequence show reduced symptoms of citrus canker disease. *J. Biotechnol.*, 167, 412–419.

Geudens, N. and Martins, J.C. (2018). Cyclic lipodepsipeptides from *Pseudomonas* spp. – Biological Swiss-Army knives. *Front. Microbiol.*, 9, 1867.

Ghag, S.B., Shekhawat, U.K.S., Ganapathi, T.R. (2012). Petunia floral defensins with unique prodomains as novel candidates for development of fusarium wilt resistance in transgenic banana plants. *PLoS One*, 7, 6.

Guan, Z.J., Guo, B., Huo, Y.L., Guan, Z.P., Dai, J.K., Wei, Y.H. (2013). Recent advances and safety issues of transgenic plant-derived vaccines. *Appl. Microbiol. Biotechnol.*, 97, 2817–2840.

Güell, I., Micaló, L., Cano, L., Badosa, E., Ferré, R., Montesinos, E., Bardají, E., Feliu, L., Planas, M. (2012). Peptidotriazoles with antimicrobial activity against bacterial and fungal plant pathogens. *Peptides*, 33, 9–17.

Guo, Q., Dong, W., Li, S., Lu, X., Wang, P., Zhang, X., Wang, Y., Ma, P. (2014). Fengycin produced by *Bacillus subtilis* NCD-2 plays a major role in biocontrol of cotton seedling damping-off disease. *Microbiol Res.*, 169(7–8), 533–540.

Gust, A., Pruitt, R., Nürnberger, T. (2017). Sensing danger: Key to activating plant immunity. *Trends Plant Sci.*, 22, 779–791.

Hancock, R.E.W. and Scott, M.G. (2000). The role of antimicrobial peptides in animal defenses. *Proc. Natl. Acad. Sci. U.S.A.*, 97, 8856–8861.

Hao, G., Zhang, S., Stover, E. (2017). Transgenic expression of antimicrobial peptide D2A21 confers resistance to diseases incited by *Pseudomonas syringae* pv. *tabaci* and *Xanthomonas citri*, but not *Candidatus* Liberibacter asiaticus. *PLoS One*, 12, 1–17.

Hegedüs, N. and Marx, F. (2013). Antifungal proteins: More than antimicrobials? *Fungal Biol. Rev.*, 26, 132–145.

Holásková, E., Galuszka, P., Mičúchová, A., Šebela, M., Öz, M.T., Frébort, I. (2018). Molecular farming in barley: Development of a novel production platform to produce human antimicrobial peptide LL-37. *Biotechnol. J.*, 13, e1700628.

Hou, S., Liu, Z., Shen, H., Wu, D. (2019). DAMP-triggered immunity in plants. *Front. Plant Sci.*, 22, 646.

Huang, C.Y., Araujo, K., Sánchez, J.N., Kund, G., Trumble, J., Roper, C., Godfrey, K.E., Jin, H. (2021). A stable antimicrobial peptide with dual functions of treating and preventing citrus Huanglongbing. *Proc. Natl. Acad. Sci. U.S.A.*, 118, 1–10.

Huffaker, A., Pearce, G., Ryan, C.A. (2006). An endogenous peptide signal in *Arabidopsis* activates components of the innate immune response. *Proc. Natl. Acad. Sci. U.S.A.*, 103, 10098–10103.

Iqbal, A., Khan, R.S., Shehryar, K., Imran, A., Ali, F., Attia, S., Shah, S., Mii, M. (2019). Antimicrobial peptides as effective tools for enhanced disease resistance in plants. *Plant Cell Tissue Organ Cult.*, 139, 1–15.

Jabrane, A., Sabri, A., Compère, P., Jacques, P., Vandenberghe, I., Van Beeumen, J., Thonart, P. (2002). Characterization of serracin P, a phage-tail-like bacteriocin, and its activity against *Erwinia amylovora*, the fire blight pathogen. *Appl. Environ. Microbiol.*, 68, 5704–5710.

Jha, S. and Chattoo, B.B. (2010). Expression of a plant defensin in rice confers resistance to fungal phytopathogens. *Transgenic Res.*, 19, 373–384.

Jung, Y.J., Lee, S.Y., Moon, Y.S., Kang, K.K. (2012). Enhanced resistance to bacterial and fungal pathogens by overexpression of a human cathelicidin antimicrobial peptide (hCAP18/LL-37) in Chinese cabbage. *Plant Biotechnol. Rep.*, 6, 39–46.

Khan, R., Najeeb, S., Hussain, S., Xie, B., Li, Y. (2020). Bioactive secondary metabolites from *Trichoderma* spp. against phytopathogenic fungi. *Microorganisms*, 8, 817.

Khosa, S., Scholz, R., Schwarz, C., Trilling, M., Hengel, H., Jaeger, K.E., Smits, S.H.J., Schmitt, L. (2017). An A/U-rich enhancer region is required for high-level protein secretion through the HlyA type I secretion system. *Appl. Environ. Microbiol.*, 15, e01163-17.

Kim, Y.S., Kim, H.M., Chang, C., Hwang, I.C., Oh, H., Ahn, J.S., Kim, K.D., Hwang, B.K., Kim, B.S. (2007). Biological evaluation of neopeptins isolated from a Streptomyces strain. *Pest Manag. Sci.*, 63, 1208–1214.

Kishi, R.N.I., Stach-Machado, D., de Lacorte Singulani, J., dos Santos, C.T., Fusco-Almeida, A.M., Cilli, E.M., Freitas-Astúa, J., Picchi, S.C., Machado, M.A. (2018). Evaluation of cytotoxicity features of antimicrobial peptides with potential to control bacterial diseases of citrus. *PLoS One*, 13, 1–18.

Kobayashi, A.K., Vieira, L.G.E., Bespalhok Filho, J.C., Leite, R.P., Pereira, L.F.P., Molinari, H.B.C., Marques, V.V. (2017). Enhanced resistance to citrus canker in transgenic sweet orange expressing the sarcotoxin IA gene. *Eur. J. Plant Pathol.*, 149, 865–873.

Koo, J.C., Chun, H.J., Park, H.C., Kim, M.C., Koo, Y.D., Koo, S.C., Ok, H.M., Park, S.J., Lee, S.H., Yun, D.J. et al. (2002). Over-expression of a seed specific hevein-like antimicrobial peptide from *Pharbitis nil* enhances resistance to a fungal pathogen in transgenic tobacco plants. *Plant Mol. Biol.*, 5, 441–452.

Kościuczuk, E.M., Lisowski, P., Jarczak, J., Strzałkowska, N., Jóźwik, A., Horbańczuk, J., Krzyżewski, J., Zwierzchowski, L., Bagnicka, E. (2012). Cathelicidins: Family of antimicrobial peptides. A review. *Mol. Biol. Rep.*, 39, 10957–10970.

Krahulec, J., Hyrsová, M., Pepeliaev, S., Jílková, J., Cerný, Z., Machálková, J. (2016). High level expression and purification of antimicrobial human cathelicidin LL-37 in *Escherichia coli. Appl. Microbiol. Biotechnol.*, 88, 167–175.

Kuddus, M.R., Rumi, F., Tsutsumi, M., Takahashi, R., Yamano, M., Kamiya, M., Kikukawa, T., Demura, M., Aizawa, T. (2016). Expression, purification and characterization of the recombinant cysteine-rich antimicrobial peptide snakin-1 in *Pichia pastoris. Protein Expr. Purif.*, 122, 15–22.

Kumar, P., Kizhakkedathu, J.N., Straus, S.K. (2018). Antimicrobial peptides: Diversity, mechanism of action and strategies to improve the activity and biocompatibility *in vivo. Biomolecules*, 8, 1.

Kumar, R., Ali, S.A., Singh, S.K., Bhushan, V., Mathur, M., Jamwal, S., Mohanty, A.K., Kaushik, J.K., Kumar, S. (2020). Antimicrobial peptides in farm animals: An updated review on its diversity, function, modes of action and therapeutic prospects. *Vet. Sci.*, 7, 1–28.

Kundu, R. (2020). Cationic amphiphilic peptides: Synthetic antimicrobial agents inspired by nature. *Chem. Med. Chem.*, 15, 1887–1896.

Kunihiro, S. and Kaneda, M. (2003). Glomecidin, a novel antifungal cyclic tetrapeptide produced by *Streptomyces lavendulae* H698 SY2. *J. Antibiot.*, 56, 30–33.

Lacerda, A.F., Del Sarto, R.P., Silva, M.S., de Vasconcelos, E.A.R., Coelho, R.R., dos Santos, V.O., Godoy, C.V., Seixas, C.D.S., da Silva, M.C.M., Grossi-de-Sa, M.F. (2016). The recombinant pea defensin Drr230a is active against impacting soybean and cotton pathogenic fungi from the genera *Fusarium*, *Colletotrichum* and *Phakopsora*. *3 Biotech.*, 6, 1–10.

Lavermicocca, P., Lonigro, S.L., Valerio, F., Evidente, A., Visconti, A. (2002). Reduction of olive knot disease by a bacteriocin from *Pseudomonas syringae* pv. *ciccaronei*. *Appl. Environ. Microbiol.*, 68, 1403–1407.

Lee, D.W. and Kim, B.S. (2021). Antimicrobial cyclic peptides for plant disease control. *Plant Pathol.*, J. 31, 1–11.

Lee, H. and Lee, D.G. (2015). Mode of action of antimicrobial peptides identified from insects. *J. Life Sci.*, 25, 715–723.

Li, L., Wang, J.X., Zhao, X.F., Kang, C.J., Liu, N., Xiang, J.H., Li, F.H., Sueda, S., Kondo, H. (2005). High level expression, purification, and characterization of the shrimp antimicrobial peptide, Ch-penaeidin, in *Pichia pastoris*. *Protein Expr. Purif.*, 39, 144–151.

Li, H.Y., Luo, Y., Zhang, X.S., Shi, W.L., Gong, Z.T., Shi, M., Chen, L.L., Chen, X.L., Zhang, Y.Z., Song, X.Y. (2015). Trichokonins from *Trichoderma pseudokoningii* SMF2 induce resistance against Gram-negative *Pectobacterium carotovorum* subsp. *carotovorum* in Chinese cabbage. *FEMS Microbiol. Lett.*, 354, 75–82.

Li, J., Hu, S., Jian, W., Xie, C., Yang X. (2021a). Plant antimicrobial peptides: Structures, functions, and applications. *Bot. Stud.*, 62, 5.

Li, A., Huang, R., Wang, C., Hu, Q., Li, H., Li, X. (2021b). Expression of anti-lipopolysaccharide factor isoform 3 in *Chlamydomonas reinhardtii* showing high antimicrobial activity. *Mar. Drugs.*, 19, 239.

Lori, M., van Verk, M., Hander, T., Schatowitz, H., Klauser, D., Flury, P., Gehring, C., Boller, T., Bartels, S. (2015). Evolutionary divergence of the plant elicitor peptides Peps produced interfamily incompatibility. *J. Exp. Bot.*, 66, 5315–5325.

Luo, Y., Zhang, D.D., Dong, X.W., Zhao, P.B., Chen, L.L., Song, X.Y., Wang, X.J., Chen, X.L., Shi, M., Zhang, Y.Z. (2010). Antimicrobial peptaibols induce defense responses and systemic resistance in tobacco against tobacco mosaic virus. *FEMS Microbiol. Lett.*, 313, 120–126.

Ma, Z., Ongena, M., Höfte, M. (2017). The cyclic lipopeptide orfamide induces systemic resistance in rice to *Cochliobolus miyabeanus* but not to *Magnaporthe oryzae*. *Plant Cell Rep.*, 36, 1731–1746.

Mansour, S.C., Pena, O.M., Hancock, R.E.W. (2014). Host defense peptides: Front-line immunomodulators. *Trends Immunol.*, 35, 443–450.

Marcos, J.F., Muñoz, A., Pérez-Payá, E., Misra, S., López-García, B. (2008). Identification and rational design of novel antimicrobial peptides for plant protection. *Annu. Rev. Phytopathol.*, 46, 273–301.

Marik, T., Tyagi, C., Balázs, D., Urbán, P., Szepesi, Á., Bakacsy, L., Endre, G., Rakk, D., Szekeres, A., Andersson, M.A. et al. (2019). Structural diversity and bioactivities of peptaibol compounds from the *Longibrachiatum* clade of the filamentous fungal genus *Trichoderma*. *Front. Microbiol.*, 10, 1434.

Mariz-Ponte, N., Regalado, L., Gimranov, E., Tassi, N., Moura, L., Gomes, P., Tavares, F., Santos, C., Teixeira, C. (2021). A synergic potential of antimicrobial peptides against *Pseudomonas syringae* pv. actinidiae. *Molecules*, 26, 5.

Maróti, G., Kereszt, A., Kondorosi, É., Mergaert, P. (2011). Natural roles of antimicrobial peptides in microbes, plants and animals. *Res. Microbiol.*, 162, 363–374.

McMillan, K.A.M. and Coombs, M.R.P. (2020). Review: Examining the natural role of amphibian antimicrobial peptide magainin. *Molecules*, 25, 5436.

Michelsen, C.F., Watrous, J., Glaring, M.A., Kersten, R., Koyama, N., Dorrestein, P.C., Stougaard, P. (2015). Nonribosomal peptides, key biocontrol components for *Pseudomonas fluorescens* In^5, isolated from a Greenlandic suppressive soil. *mBio*, 6, e00079-15.

Mirzaee, M., Holásková, E., Mičúchová, A., Kopečný, D.J., Osmani, Z., Frébort, I. (2021). Long-lasting stable expression of human LL-37 antimicrobial peptide in transgenic barley plants. *Antibiotics*, 10, 898.

Mohan, S., Meiyalaghan, S., Latimer, J.M., Gatehouse, M.L., Monaghan, K.S., Vanga, B.R., Pitman, A.R., Jones, E.E., Conner, A.J., Jacobs, J.M.E. (2014). GSL2 over-expression confers resistance to *Pectobacterium atrosepticum* in potato. *Theor. Appl. Genet.*, 127, 677–689.

Moll, L., Badosa, E., Planas, M., Feliu, L., Montesinos, E., Bonaterra, A. (2021). Antimicrobial peptides with antibiofilm activity against *Xylella fastidiosa*. *Front. Microbiol.*, 12, 753874.

Monroc, S., Badosa, E., Feliu, L., Planas, M., Montesinos, E., Bardají, E. (2006). De novo designed cyclic cationic peptides as inhibitors of plant pathogenic bacteria. *Peptides*, 27, 2567–2574.

Montesinos, E. (2007). Antimicrobial peptides and plant disease control. *FEMS Microbiol. Lett.*, 270, 1–11.

Montesinos, E., Badosa, E., Cabrefiga, J., Planas, M., Feliu, L., Bardají, E. (2012). Antimicrobial peptides for plant disease control. From discovery to application. In *Small Wonders: Peptides for Disease Control*, Rajasekaran, K., Cary, J., Jaynes, J., Montesinos, E. (eds). Oxford University Press, Oxford.

Montesinos, L., Bundó, M., Izquierdo, E., Campo, S., Badosa, E., Rossignol, M., Montesinos, E., San Segundo, B., Coca, M. (2016). Production of biologically active cecropin a peptide in rice seed oil bodies. *PLoS One*, 11, e0146919.

Montesinos, L., Bundó, M., Badosa, E., San Segundo, B., Coca, M., Montesinos, E. (2017). Production of BP178, a derivative of the synthetic antibacterial peptide BP100, in the rice seed endosperm. *BMC Plant Biol.*, 17, 63.

Montesinos, L., Gascón, B., Ruz, L., Badosa, E., Planas, M., Feliu, L., Montesinos, E. (2021). A bifunctional synthetic peptide with antimicrobial and plant elicitation properties that protect tomato plants from bacterial and fungal infections. *Front. Plant Sci.*, 12, 756357.

Mora, I., Cabrefiga, J., Montesinos, E. (2015). Cyclic lipopeptide biosynthetic genes and products, and inhibitory activity of plant-associated *Bacillus* against phytopathogenic bacteria. *PLoS One*, 10, e0127738.

Mouloud, G., Daoud, H., Bassem, J., Laribi Atef, I., Hani, B. (2013). New bacteriocin from *Bacillus clausii* strainGM17: Purification, characterization, and biological activity. *Appl. Biochem. Biotechnol.*, 171(8), 2186–2200.

Muñoz, A., Gandía, M., Harries, E., Carmona, L., Read, D., Marcos J.F. (2013). Understanding the mechanism of action of cell-penetrating antifungal peptides using the rationally designed hexapeptide PAF26 as a model. *Fungal. Biol. Rev.*, 26, 146–155.

Nadal, A., Montero, M., Company, N., Badosa, E., Messeguer, J., Montesinos, L., Montesinos, E., Pla, M. (2012). Constitutive expression of transgenes encoding derivatives of the synthetic antimicrobial peptide BP100: Impact on rice host plant fitness. *BMC Plant Biol.*, 12, 159

Nes, I.F., Diep, D.B., Havarstein, L.S., Brurberg, M.B., Eijsink, V., Holo, H. (1996). Biosynthesis of bacteriocins in lactic acid bacteria. *Antonie van Leeuwenhoek*, 70, 113–128.

Ng-Choi, I., Soler, M., Güell, I., Badosa, E., Cabrefiga, J., Bardají, E., Montesinos, E., Planas, M., Feliu, L. (2014). Antimicrobial peptides incorporating non-natural amino acids as agents for plant protection. *Protein Pept. Lett.*, 21, 4.

Nielsen, T.H., Thrane, C., Christophersen, C., Anthoni, U., Sorensen, J. (2000). Structure, production characteristics and fungal antagonism of Tensin – A new antifungal cyclic lipopeptide from *Pseudomonas fluorescens* strain 96–578. *J. Appl. Microbiol.*, 898, 992–1001.

Oliveras, À., Baró, A., Montesinos, L., Badosa, E., Montesinos, E., Feliu, L., Planas, M. (2018). Antimicrobial activity of linear lipopeptides derived from BP100 towards plant pathogens. *PLoS One*, 13, 1–17.

Oliveras, A., Moll, L., Riesco-Llach, G., Tolosa-Canudas, A., Gil-Caballero, S., Badosa, E., Bonaterra, A., Montesinos, E., Planas, M., Feliu, L. (2021). D-amino acid-containing lipopeptides derived from the lead peptide BP100 with activity against plant pathogens. *Int. J. Mol. Sci.*, 22, 12.

Ongena, M., Jacques, P., Touré, Y., Destain, J., Jabrane, A., Thonart, P. (2005). Involvement of fengycin-type lipopeptides in the multifaceted biocontrol potential of *Bacillus subtilis*. *Appl. Microbiol. Biotechnol.*, 69, 29–38.

Osada, H. (2019). Discovery and applications of nucleoside antibiotics beyond polyoxin. *J. Antibiot.*, 72, 855–864.

Parachin, N.S., Mulder, K.C., Viana, A.A., Dias, S.C., Franco O.L. (2012). Expression systems for heterologous production of antimicrobial peptides. *Peptides*, 38, 446–456.

Preecha, C., Sadowsky, M.J., Prathuangwong, S. (2010). Lipopeptide surfactin produced by *Bacillus amyloliquefaciens* KPS46 is required for biocontrol efficacy against *Xanthomonas axonopodis* pv. glycines. *Kasetsart J. (Nat. Sci.)*, 44, 84–99.

Príncipe, A., Fernandez, M., Torasso, M., Godino, A., Fischer, S. (2018). Effectiveness of tailocins produced by *Pseudomonas fluorescens* SF4c in controlling the bacterial-spot disease in tomatoes caused by *Xanthomonas vesicatoria*. *Microbiol. Res.*, 212–213, 94–102.

Puig, M., Moragrega, C., Ruz, L., Montesinos, E., Llorente, I. (2015). Controlling brown spot of pear by a synthetic antimicrobial peptide under field conditions. *Plant Dis.*, 99, 1816–1822.

Raaijmakers, J.M., De Bruijn, I., Nybroe, O., Ongena, M. (2010). Natural functions of lipopeptides from *Bacillus* and *Pseudomonas*: More than surfactants and antibiotics. *FEMS Microbiol. Rev.*, 34, 1037–1062.

Rahnamaeian, M. and Vilcinskas, A. (2012). Defense gene expression is potentiated in transgenic barley expressing antifungal peptide metchnikowin throughout powdery mildew challenge. *J. Plant Prot. Res.*, 125, 115–124.

Ramadevi, R., Rao, K.V., Reddy, V.D. (2014). *Agrobacterium tumefaciens*-mediated genetic transformation and production of stable transgenic pearl millet (*Pennisetum glaucum* [L.] R. Br.). *In Vitro Cell. Dev. Biol. – Plant.*, 50, 392–400.

Rhee, K.H. (2003). Purification and identification of an antifungal agent from *Streptomyces* sp. KH-614 antagonistic to rice blast fungus, *Pyricularia oryzae*. *J. Mol. Microbiol. Biotechnol.*, 13, 984–988.

Romero, D., de Vicente, A., Rakotoaly, R.H., Dufour, S.E., Veening, J.W., Arrebola, E., Cazorla, F.M., Kuipers, O.P., Paquot, M., Pérez-García, A. (2007). The iturin and fengycin families of lipopeptides are key factors in antagonism of *Bacillus subtilis* toward *Podosphaera fusca*. *Mol. Plant Microbe Interact.*, 20, 430–440.

Ross, A., Yamada, K., Hiruma, K., Yamashita-Yamada, M., Lu, X., Takano, Y., Tsuda, K, Saijo, Y. (2014). The *Arabidopsis* PEPR pathway couples local and systemic plant immunity. *The EMBO J.*, 33, 62–75.

Rufo, R., Batlle, A., Camprubi, A., Montesinos, E., Calvet, C. (2017). Control of Rubus stunt and stolbur diseases in Madagascar periwinkle with mycorrhizae and a synthetic antibacterial peptide. *Plant Pathol.*, 66, 551–558.

Ruiz, C., Nadal, A., Montesinos, E., Pla, M. (2018a). Novel Rosaceae plant elicitor peptides as sustainable tools to control *Xanthomonas arboricola* pv. pruni in *Prunus* spp. *Mol. Plant Pathol.*, 19, 418–431.

Ruiz, C., Nadal, A., Foix, L., Montesinos, L., Montesinos, E., Pla, M. (2018b). Diversity of plant defense elicitor peptides within the *Rosaceae*. *BMC Genet.*, 19, 11.

Sarwar, A., Hassan, M.N., Imran, M., Iqbal, M., Majeed, S., Brader, G., Sessitsch, A., Hafeez, F.Y. (2018). Biocontrol activity of surfactin A purified from *Bacillus* NH-100 and NH-217 against rice bakanae disease. *Microbiol. Res.*, 209, 1–13.

Schaefer, S.C., Gasic, K., Cammue, B., Broekaert, W., van Damme, E.J., Peumans, W.J., Korban, S.S. (2005). Enhanced resistance to early blight in transgenic tomato lines expressing heterologous plant defense genes. *Planta*, 222, 858–866.

Scholz, R., Vater, J., Budiharjo, A., Wang, Z., He, Y., Dietel, K., Schwecke, T., Herfort, S., Lasch, P., Borriss, R. (2014). Amylocyclicin, a novel circular bacteriocin produced by *Bacillus amyloliquefaciens* FZB42. *J. Bacteriol.*, 196, 1842–1852.

Schubert, M., Houdelet, M., Kogel, K.H., Fischer, R., Schillberg, S., Nölke, G. (2015). Thanatin confers partial resistance against aflatoxigenic fungi in maize (*Zea mays*). *Transgenic Res.*, 24, 885–895.

Shanmugaraj, B., Bulaon, C.J.I., Malla, A., Phoolcharoen, W. (2021). Biotechnological insights on the expression and production of antimicrobial peptides in plants. *Molecules*, 26, 4032.

Sharma, A.K. and Sharma, M.K. (2009). Plants as bioreactors: Recent developments and emerging opportunities. *Biotechnol. Adv.*, 27, 811–832.

Sher Khan, R., Iqbal, A., Malak, R., Shehryar, K., Attia, S., Ahmed, T., Ali Khan, M., Arif, M., Mii, M. (2019). Plant defensins: Types, mechanism of action and prospects of genetic engineering for enhanced disease resistance in plants. *3 Biotech.*, 9, 1–12.

Shi, M., Chen, L., Wang, X.W., Zhang, T., Zhao, P.B., Song, X.Y., Sun, C.Y., Chen, X.L., Zhou, B.C., Zhang, Y.Z. (2012). Antimicrobial peptaibols from *Trichoderma pseudokoningii* induce programmed cell death in plant fungal pathogens. *Microbiology*, 158, 166–175.

Smits, T.H.M., Duffy, B., Blom, J., Ishimaru, C.A., Stockwell, V.O. (2019). Pantocin A, a peptide-derived antibiotic involved in biological control by plant-associated *Pantoea* species. *Arch. Microbiol.*, 201, 713–722.

Stein, T. (2005). *Bacillus subtilis* antibiotics: Structures, syntheses and specific functions. *Mol. Microbiol.*, 56, 845–857.

Stein, T., Borchert, S., Conrad, B., Feesche, J., Hofemeister, B., Hofemeister, J., Entian, K.D. (2002). Two different lantibiotic-like peptides originate from the ericin gene cluster of *Bacillus subtilis* A1/3. *J. Bacteriol.*, 184(6), 1703–1711.

Stover, E.D., Stange, R.R., Mccollum, T.G., Jaynes, J., Irey, M., Mirkov, E. (2013). Screening antimicrobial peptides in vitro for use in developing transgenic *Citrus* resistant to huanglongbing and citrus canker. *J. Am. Soc. Hortic. Sci.*, 138, 142–148.

Subramanian, S. and Smith, D.L. (2015). Bacteriocins from the rhizosphere microbiome – From an agriculture perspective. *Front. Plant Sci.*, 6, 909.

Tam, J., Wang, S., Wong, K., Tan, W.L. (2015). Antimicrobial peptides from plants. *Pharmaceuticals*, 8, 711–757.

Tran, H., Ficke, A., Asiimwe, T., Hofte, M., Raaijmakers, J.M. (2007). Role of the cyclic lipopeptide massetolide A in biological control of *Phytophthora infestans* and in colonization of tomato plants by *Pseudomonas fluorescens*. *New Phytol.*, 175, 731–742.

Ugras, S., Sezen, K., Kati, H., Demirbag, Z. (2013). Purification and characterization of the bacteriocin Thuricin Bn1 produced by *Bacillus thuringiensis* subsp. *kurstaki* Bn1 isolated from a hazelnut pest. *J. Microbiol. Biotechnol.*, 23, 167–176.

Vila, L., Lacadena, V., Fontanet, P., Martinez del Pozo, A., San Segundo, B. (2001). A protein from the mold *Aspergillus giganteus* is a potent inhibitor of fungal plant pathogens. *Mol. Plant Microbe Interact.*, 14, 1327–1331.

Viterbo, A., Wiest, A., Brotman, Y., Chet, I., Kenerley, C. (2007). The 18 mer peptaibols from *Trichoderma virens* elicit plant defence responses. *Mol. Plant Pathol.*, 8, 737–746.

Wang, Y.Q. and Cai, J.Y. (2007). High-level expression of acidic partner-mediated antimicrobial peptide from tandem genes in *Escherichia coli*. *Appl. Biochem. Biotechnol.*, 141, 203–213.

Wu, Q., Patočka, J., Kuča, K. (2018). Insect antimicrobial peptides, a mini review. *Toxins*, 10, 1–17.

Yun, B.S., Yoo, I.D., Kim, Y.H., Kim, Y.S., Lee, S.J., Kim, K.S., Yeo, W.H. (2000). Peptaivirins A and B, two new antiviral peptaibols against TMV infection. *Tetrahedron Lett.*, 41, 1429–1431.

Zachow, C., Jahanshah, G., de Bruijn, I., Song, C., Ianni, F., Pataj, Z., Gerhardt, H., Pianet, I., Lämmerhofer, M., Berg, G. et al. (2015). The novel lipopeptide poaeamide of the endophyte *Pseudomonas poae* RE*1-1-14 is involved in pathogen suppression and root colonization. *Mol. Plant Microbe Interact.*, 28, 800–810.

Zeitler, B., Bernhard, A., Meyer, H., Sattler, M., Koop, H.-U., Lindermayr, C. (2013). Production of a de-novo designed antimicrobial peptide in *Nicotiana benthamiana*. *Plant Mol. Biol.*, 81, 259–272.

Zhang, D., Lu, Y., Chen, H., Wu, C., Zhang, H., Chen, L., Chen, X. (2020). Antifungal peptides produced by actinomycetes and their biological activities against plant diseases. *J. Antibiot.*, 73, 265–282.

Zhou, M., Hu, Q., Li, Z., Li, D., Chen, C.F., Luo, H. (2011). Expression of a novel antimicrobial peptide Penaeidin4-1 in creeping bentgrass (*Agrostis stolonifera* L.) enhances plant fungal disease resistance. *PLoS One*, 6, 8–10.

Zhu, X., Li, Z., Xu, H., Zhou, M., Du, L., Zhang, Z. (2012). Overexpression of wheat lipid transfer protein gene TaLTP5 increases resistances to *Cochliobolus sativus* and *Fusarium graminearum* in transgenic wheat. *Funct. Integr. Genomics*, 12, 481–488.

7

Biocontrol of Plant Pathogens via Quorum Quenching

Denis FAURE[1] and Xavier LATOUR[2]
[1] I2BC, Paris-Saclay University, France
[2] CBSA UR4312, Normandy University, France

In several bacterial pathogens, quorum sensing signaling controls the expression of virulence factors. Quorum quenching deals with the identification and use of chemical and biological agents to disturb quorum sensing in pathogens. This chapter will highlight the quorum quenching strategy for protecting plants using biocontrol agents, and will discuss the advantages and constraints of this approach as well as its potential use in combination with other biocontrol strategies.

7.1. Quorum quenching to counteract quorum sensing

Quorum sensing (QS) is a gene regulation mechanism that associates the synthesis and perception of small diffusible molecules with the control of gene expression in bacteria (Fuqua et al. 1994). Some bacteria synthesize signal molecules, called auto-inducers or QS signals that diffuse outside/inside the bacterial cells. QS signals accumulate at a threshold concentration, which allows the binding of regulatory proteins to, in turn, modulate – activate or repress – gene transcription. Two main conditions facilitate the accumulation of QS signals: (i) a high proliferation of the QS-emitting population, hence an increase in QS signal production, and (ii) a

Biocontrol of Plant Disease,
coordinated by Claire PRIGENT-COMBARET and Bernard DUMAS. © ISTE Ltd 2022.

decrease of mass transfer, hence a lower dilution of QS signals in the environment.

The chemical structure of QS signals is diverse: for instance, *N*-acylhomoserine lactones (AHLs) in a wide diversity of Proteobacteria, oligopeptides in Firmicutes and cis-11-methyl-2-dodecenoic acid (called the DSF for Diffusible Signal Factor) in several *Xanthomonas* species (see a review by Grandclément et al. 2016). In numerous pathogens, QS controls the expression of genes involved in the virulence, colonization and exploitation of a plant host, as well as the dissemination of the virulence gene by horizontal transfer (Baltenneck et al. 2021). In the necrotrophic pathogen *Pectobacterium atrosepticum*, the QS signal *N*-3-oxo-octanoylhomoserinelactone (OC8HSL) regulates the synthesis of proteases and pectinolytic enzymes involved in the degradation of the plant cell wall (Baltenneck et al. 2021). In the biotrophic pathogen *Agrobacterium tumefaciens*, the same signal OC8HSL controls the conjugation of the virulence Ti plasmid (Lang and Faure 2014). In *Xanthomonas campestris*, the DSF regulates the synthesis of exopolysaccharides, cellulases and proteases (He and Zhang 2008).

Quorum quenching (QQ) refers to any processes that disrupt QS and hence abolish the regulation of QS-regulated traits (Dong et al. 2001). QQ may impair different steps of the QS regulatory pathway, such as QS signal synthesis, QS signal accumulation or QS signal interaction with regulatory proteins. A wide spectrum of natural and synthetic chemicals exhibit QQ activity: they are called QS inhibitors (QSIs). Some enzymes such as lactonases and amidases also exhibit QQ activity because they cleave or modify QS signals. All these aspects were reviewed by Grandclément et al. (2016). Some living organisms expressing QQ activities and their natural active components (QSIs and QQ enzymes) have been proposed to be used as biocontrol agents for impairing QS-regulated traits of pathogens.

7.2. Quorum sensing inhibitors

Numerous and extensive QSI screens were performed on animal and human pathogens, such as *Pseudomonas aeruginosa*, but less on plant pathogens (Defoirdt 2018). Noticeably, some QSIs inhibit the growth of pathogens; hence, QQ and bactericidal and bacteriostatic activities could not be clearly discriminated in some studies. Important efforts were made to identify QSIs against *Pectobacterium* species that infect a large spectrum of plants. Salicylic acid seems to impair QS of *Pectobacterium parmentieri* because of its binding to the enzyme ExpI that is involved in the QS signal synthesis (Joshi et al. 2020). Another plant defense compound, phloretin, a phytoalexin from apple, also seems to interfere with ExpI and QS signal synthesis in *Pectobacterium brasiliense* (Pun et al. 2021).

Some other compounds, such as halogenated furanones from the red alga, which target the QS regulator LuxR in *Vibrio fischeri* and its homologs, were able to interfere with the QS of *Pectobacterium* strains (Manefield et al. 2001; Koch et al. 2005). Hexanal, a food additive, and piericidin A and glucopiericidin A, which are produced by the bacterium *Streptomyces xanthocidicus*, as well as the synthetic *N,N'*-bisalkylated imidazolium salts exhibit QQ activity against *Pectobacterium* strains (Raoul des Essarts et al. 2013; Kang et al. 2016; Zhang et al. 2018); the mechanism of action of these QSIs is unknown.

QSIs were discovered to target other plant pathogens: Chumacin-1 and Chumacin-2 from *Pseudomonas aeruginosa* are able to block the DSF QS signaling in *Xanthomonas oryzae* pv. *oryzae* (Kanugala et al. 2019); methyl 3-hydroxy-8-phenyloctanoate and its synthetic derivatives inhibit QS using (R)-methyl 3-hydroxymyristate (3-OH PAME) in *Ralstonia solanacearum* (Yoshihara et al. 2020); drimendiol (a sesquiterpene) acts as a QSI against *Pseudomonas syringae* and increases the biocide effects of $CuSO_4$ on *P. syringae* biofilms (Paza et al. 2013).

7.3. Quorum quenching enzymes

QQ enzymes alter the structure of QS signals, thereby impairing QS signaling. These enzymes are diverse because the structure of substrates, the QS signals (e.g. AHLs and DSF), are diverse, as well as the different enzymatic activities (e.g. lactonase and amidase) that may act on the same QS signal. These enzymatic types were reviewed by Grandclément et al. (2016). In the case of AHLs, two main enzymatic activities were characterized: lactonases open the lactone ring, while amidases cleave the amide bond and release homoserine lactone and fatty acid. Two additional enzymatic activities were reported as acting on these QS signals: reductases that convert 3-oxo-substituted into 3-hydroxy-substituted QS signals and oxidases that modify the acyl chain. The QS signals 3-OH PAME from *R. solanacearum* may be altered by specific esterases (Shinohara et al. 2007).

QQ enzymes were present in a wide diversity of eubacteria and eukaryotes (including mammals), as well as in some archaea. Some QQ enzymes are produced by the bacteria that emit QS signals themselves such as lactonases in *A. tumefaciens* and amidases in *P. syringae*. In these cases, they could contribute to the regulation of the QS signal pool in pathogens (Shepherd and Lindow 2009; Dessaux and Faure 2018). Some plants exhibit QQ lactonase activity, but the genetic determinant is not known (Delalande et al. 2005). Bacterial genes coding for QQ lactonases were transferred into plants to decrease their sensitivity to *Pectobacterium* strains (Dong et al. 2001). Aside this approach using genetically modified plants, a promising approach could be the use of a hyperthermostable lactonase purified from the archaeon *Sulfolobus solfataricus* (Elias et al. 2008).

7.4. Quorum quenching biocontrol agents

QQ activity may be considered as a trait to be searched for in identifying biocontrol agents (Molina et al. 2003). For almost 20 years, dozens of papers have reported the screening and isolation of microorganisms exhibiting QQ activity (QSI or QQ enzyme) and biocontrol activity simultaneously, but the genes involved in QQ activity remain often unknown (for instance Rodriguez et al. 2020; Zhang et al. 2021).

A remarkable discovery was the presence of the *aiiA* gene coding for a QQ lactonase in *Bacillus thuringiensis*, a well-known biocontrol agent against insects (Dong et al. 2004). A constructed *aiiA* KO-mutant was defective for lactonase activity against the QS-AHL signal. The *aiiA* KO-mutant was also impaired in the protection of potato tubers infected by a *Pectobacterium* strain. The QQ activity was reported in several *Bacillus* species, which could be used as biocontrol agents against different plant and animal (e.g. fish) pathogens expressing QS (Dong et al. 2004; Chu et al. 2017).

Another noticeable bacterial species is *Rhodococcus erythropolis*, a common inhabitant of soil and rhizosphere. This species has expressed several QQ activities – lactonase, amidase and reductase – and has protected potato tubers against *Pectobacterium* strains (Uroz et al. 2005, 2008). Genome and transcriptome of the *R. erythropolis* strain R138 were analyzed (Kwasiborski et al. 2015a, 2015b). The gene *qsdA* coding for a large-spectrum QQ lactonase was characterized (Uroz et al. 2008). The gene *qsdA* belongs to an operon under the control of the transcriptional regulator

QsdR (El Sahili et al. 2015; Barbey et al. 2018). The expression of the *qsd* operon is activated in the presence of AHLs and structural analogs (Barbey et al. 2018; Chane et al. 2019a), as well as in potato tubers in the presence of the pathogen *Pectobacterium atrosepticum* (Barbey et al. 2013; Chane et al. 2019b).

7.5. Monitoring of quorum quenching biocontrol agents and activities

The use of promoter-probe vectors carrying fluorescent protein reporter genes appears to be a relevant way to study both the epidemiology and etiology of phytopathogenic and biocontrol agents in situ. This technology coupled with adequate microscopy has made it possible, additionally to verify bacterial viability, to provide direct visual evidence of colonization kinetics on the host plant, as well as the capacity of each agent to form microcolonies and/or biofilms (Bourigault et al. 2021a). In *R. erythropolis* strain R138, the recent understanding of the regulatory mechanism of *qsd* operon expression (see section 7.4) made it possible to remove a fundamental technological lock in order to construct the first biosensor of AHL-based QQ activity (Barbey et al. 2018). A rhodococcal dual-colored reporter was constructed : while the mCherry fluorophore is expressed as a cell tag, the green fluorescent protein (GFP) was used as a regulator-based reporter system to simultaneously visualize in planta bacterial spread and QQ activity leading to the biocontrol of potato soft-rot (Chane et al. 2019b and 2020). This revealed the chronology of events leading to tuber maceration and QS communication, in addition to the disruption of the latter and concomitant plant protection by the rhodococcal biocontrol agent.

The same *Rhodococcus* QQ biosensor was more recently used to sense both QS and QQ activities within a large biofilm made by *Rhizobium rhizogenes* (formerly *Agrobacterium rhizogenes*), a canker pathogen known to generate hairy root disease in hydroponic crops (Abarca-Grau et al. 2011; Bosmans et al. 2015). Here, AHL catabolism was particularly observed inside the regions containing a mixture of large amounts of rhizobial and rhodococcal cells, a proximity of partners necessary for AHL penetration inside rhodococci to be degraded (Bourigault et al. 2021b). Interestingly, the QQ activity of the R138 biocontrol agent resulted in the modification of the characteristics of the biofilm formed by *R. rhizogenes*, such as the alteration of biomass and biovolume as well as the structure of the biofilm. These alterations are related to the disruption of QS communication deployed by rhizobia in the biofilm, probably because the jamming of AHL communication has a drastic effect on the metabolism of rhizobial exopolysaccharides (Abarca-Grau et al. 2011; Bourigault et al. 2021b).

In addition to the QQ bacteria, the activity of isolated QQ enzymes can also be monitored in situ. For example, the functions of AHL-lactonase QsdA were first investigated by transferring the *qsdA* gene to a heterologous host (*Escherichia coli*), thus becoming a powerful AHL quencher (Barbey et al. 2013). Then, the lactonase QsdA was overproduced and purified to prevent the formation of the *R. rhizogenes* biofilm (Bourigault et al. 2021b). In this study, the QsdA lactonase, even isolated from its progenitor, shows major anti-biofilm efficacy, similar to that recorded with the R138 strain or the heterologous host *E. coli*, under assay conditions. This confirms the significant QQ potential already observed in other AHL lactonases of the phosphotriesterase family to which QsdA belongs (Elias et al. 2008; Guendouze et al. 2017).

7.6. Biostimulation of quorum quenching

QQ biostimulation refers to the processes that stimulate the proliferation of native or introduced QQ populations, thereby contributing to a global increase of QQ activities in the biostimulated community (Cirou et al. 2007, 2012). A screen of different lactones revealed treatments with gamma-caprolactone and gamma-heptanolactone as efficient for increasing the relative abundance of native and introduced (*R. erythropolis*) QQ populations in the hydroponic culture of potato plants (Cirou et al. 2011, 2012). The two lactones were assimilated as the carbon source by QQ populations, and therefore did not accumulate in plant tissues or a hydroponic culture medium. The treatment with these lactones could facilitate the settlement of introduced QQ populations in the potato rhizosphere. Remarkably, the efficiency of QQ biostimulation with different treatments (including with gamma-caprolactone) was demonstrated to control biofouling in the membrane bioreactor for wastewater treatment (Yu et al. 2021).

In order to assess the impact of these treatments on the structure of bacterial communities in the rhizosphere, the analysis of bacterial diversity by 454-Roche sequencing of amplicons of the 16S ribosomal gene was performed (Cirou et al. 2012). The overall structure of the communities of the samples treated with the coupling approach associating gamma-heptanolactone and *R. erythropolis* R138 is different from control samples. The level of the target population (bacteria belonging to the genus *Rhodococcus*) is most increased in the coupling treatment. Using the 454-ARN16S approach, this data consolidates the achievements of cultivable approaches. Among the more abundant genera, that is, whose representation is greater than 1/1,000 in communities, dominant genera such as *Pseudomonas* or *Acidovorax* are not affected by the treatments, unlike target populations such as *Rhodococcus* (Cirou et al. 2012). Taken together, this data suggests that the treatments have a measurable impact, but do not affect the dominant populations of the rhizosphere.

7.7. Management of quorum quenching treatments

The reliability of biocontrol practices has often generated skepticism because of a lack of consistency in disease control, due to an insufficient rhizosphere competence and activity of biocontrol agents (Compant et al. 2005; Weller 2007; Latour et al. 2009). Therefore, an adequate way to improve the effects of biocontrol may require a crop management referred to as "rhizosphere engineering" (Ryan et al. 2009; Berg et al. 2013). Rhizosphere engineering consists of amending the soil of engineering plants and/or soil microorganisms, including the interactions between partners, to promote plant growth and health (Dessaux et al. 2016). This type of management should also be implemented for QQ control procedures. Indeed, an efficient disease control using a QQ-based biocontrol needs a potent AHL signal degrading activity operated by the biocontrol candidate and therefore the full expression of genes encoding for QQ enzymes.

As revealed previously, the gamma-caprolactone appears as a model of promising and powerful biostimulating amendments for QQ-based biocontrol. Indeed, it can favor a biocontrol by conservation with a transient elevation of native bacterial communities, which are able to metabolize both biostimulating and AHL molecules in a non-specific manner (Cirou et al. 2007, 2011). Gamma-caprolactone can also be spread with a rhodococcal biocontrol agent to sustain its fitness in the plant vicinity supporting a biocontrol by augmentation (Cirou et al. 2012; Latour et al. 2013). Accordingly, the relevant use of this compound will require special formulation conditions, which should go beyond the single dissolution in a suitable liquid solvent. Indeed, to be fully active, QQ biostimulating molecules will have to be incorporated into a sustainable formulation applicable for a large-scale spreading. Since rhodococci are bacteria that can be readily incorporated into formulations (Vancov et al. 2006; Kuyukina et al. 2013; Kim et al. 2015), it is contemplated that a formulation of the *R. erythropolis* strain R138 may be bagged in a biodegradable polycaprolactone film, and then spread during potato planting, for example. This should allow the prolonged release of the biocontrol agent at the same time as that of caprolactone monomers, promoting both rhodococcal growth and quenching activity throughout the plant's growth cycle (Latour et al. 2013; Chane et al. 2019a).

7.8. Quorum quenching in biocontrol: perspectives

Over the last two decades, QQ has emerged as a novel biocontrol trait for protecting plants and animals against QS-regulated functions of pathogens. QQ molecular actors (mainly QSIs and QQ enzymes) are diverse and accessible in plant environments: up to 5% of the culturable bacteria exhibit QQ activity (d'Angelo-Picard et al. 2005; Cirou et al. 2012). In the near future, additional screens of natural compounds and microorganisms could increase the panel of bio-sourced QSIs and biocontrol agents against plant pathogens. Biostimulation could contribute to improving the efficiency of biocontrol treatments. Importantly, QQ limits the expression of QS-regulated functions, but does not impair the proliferation of pathogens. Hence, pathogens could remain persistent even upon QQ treatments. To circumvent this problem, a combination of QQ and antibiosis approaches may be proposed to fight the virulence and proliferation of pathogens. Another important issue is that several biocontrol agents use QS to regulate beneficial traits: for instance, phenazine production is regulated by QS in the biocontrol agent *Pseudomonas chlororaphis* (Molina et al. 2003). A combination of treatments could take into account the potential interferences between QQ and QS traits of biocontrol agents (Molina et al. 2003). Field assays are now important to validate QQ biocontrol efficiency and its combination with other biocontrol strategies.

7.9. Acknowledgments

DF thanks CNRS, University Paris-Saclay and LabEx Saclay Plant Sciences-SPS (ANR-10-LABX-0040-SPS) for financial support. XL thanks the Région Normandie, FEDER (U.E.) and the Evreux Portes de Normandie agglomération (RIN BEER and Norman Crops Biocontrol projects).

7.10. References

Abarca-Grau, A.M., Penyalver, R., López, M.M., Marco-Noales, E. (2011). Pathogenic and non-pathogenic *Agrobacterium tumefaciens*, *A. rhizogenes* and *A. vitis* strains form biofilms on abiotic as well as on root surfaces. *Plant Pathol.*, 60, 416–425.

d'Angelo-Picard, C., Faure, D., Penot, I., Dessaux, Y. (2005). Diversity of N-acyl homoserine lactone-producing and -degrading bacteria in soil and tobacco rhizosphere. *Environ. Microbiol.*, 7(11), 1796–1808.

Baltenneck, J., Reverchon, S., Hommais, F. (2021). Quorum sensing regulation in phytopathogenic bacteria. *Microorganisms*, 9(2), 239.

Barbey, C., Crépin, A., Bergeau, D., Ouchiha, A., Mijouin, L., Taupin, L., Orange, N., Feuilloley, M., Dufour, A., Burini, J.F. et al. (2013). In planta biocontrol of *Pectobacterium atrosepticum* by *Rhodococcus erythropolis* involves silencing of pathogen communication by the Rhodococcal gamma-lactone catabolic pathway. *PLoS One*, 8(6), e66642.

Barbey, C., Chane, A., Burini, J.F., Maillot, O., Merieau, A., Gallique, M., Beury-Cirou, A., Konto-Ghiorghi, Y., Feuilloley, M., Gobert, V. et al. (2018). A Rhodococcal transcriptional regulatory mechanism detects the common lactone ring of AHL quorum-sensing signals and triggers the quorum-quenching response. *Front. Microbiol.*, 9, 2800.

Berg, G., Zachow, C., Müller, H., Philipps, J., Tilcher, R. (2013). Next-generation bio-products sowing the seeds of success for sustainable agriculture. *Agronomy*, 3, 648–656.

Bosmans, L., Álvarez-Pérez, S., Moerkens, R., Wittemans, L., Van Calenberge, B., Kerckhove, S.V., Paeleman, A., De Mot, R., Rediers, H., Lievens, B. (2015) Assessment of the genetic and phenotypic diversity among rhizogenic *Agrobacterium* biovar 1 strains infecting solanaceous and cucurbit crops. *FEMS Microbiol. Ecol.*, 91, fiv081.

Bourigault, Y., Chane, A., Barbey, C., Jafra, S., Czajkowski, R., Latour, X. (2021a). Biosensors used for epifluorescence and confocal laser scanning microscopies to study *Dickeya* and *Pectobacterium* virulence and biocontrol. *Microorganisms*, 9, 295.

Bourigault,Y., Rodrigues, S., Crépin, A., Chane, A., Taupin, L., Bouteiller, M., Dupont, C., Merieau, A., Konto-Ghiorghi, Y., Boukerb, A.M. et al. (2021b). Biocontrol of biofilm formation: Jamming of sessile-associated rhizobial communication by rhodococcal quorum-quenching. *Int. J. Mol. Sci.*, 22, 8241.

Chane, A., Barbey, C., Bourigault, Y., Maillot, O., Rodrigues, S., Bouteiller, M., Merieau, A., Konto-Ghiorghi, Y., Beury-Cirou, A., Gattin, R. et al. (2019a). A flavor lactone mimicking AHL quorum-sensing signals exploits the broad affinity of the QsdR regulator to stimulate transcription of the Rhodococcal *qsd* operon involved in quorum-quenching and biocontrol activities. *Front. Microbiol.*, 10, 786.

Chane, A., Barbey, C., Robert, M., Merieau, A., Konto-Ghiorghi, Y., Beury-Cirou, A., Feuilloley, M., Pátek, M., Gobert, V., Latour, X. (2019b). Biocontrol of soft rot: Confocal microscopy highlights virulent pectobacterial communication and its jamming by rhodococcal quorum-quenching. *Mol. Plant Microbe Interact.*, 32(7), 802–812.

Chane, A., Bourigault, Y., Bouteiller, M., Konto-Ghiorghi, Y., Merieau, A., Barbey, C., Latour, X. (2020). Close-up on a bacterial informational war in the geocaulosphere. *Can. J. Microbiol.*, 66, 447–454.

Chu, W., Zhou, S., Zhu, W., Zhuang, X. (2017). Quorum quenching bacteria *Bacillus* sp. QSI-1 protect zebrafish (*Danio rerio*) from *Aeromonas hydrophila* infection. *Sci. Rep.*, 4, 5446.

Cirou, A., Diallo, S., Kurt, C., Latour, X., Faure, D. (2007). Growth promotion of quorum-quenching bacteria in the rhizosphere of *Solanum tuberosum*. *Environ. Microbiol.*, 9(6), 1511–1522.

Cirou, A., Raffoux, A., Diallo, S., Latour, X., Dessaux, Y., Faure, D. (2011). Gamma-caprolactone stimulates growth of quorum-quenching *Rhodococcus* populations in a large-scale hydroponic system for culturing *Solanum tuberosum*. *Res. Microbiol.*, 162(9), 945–950.

Cirou, A., Mondy, S., An, S., Charrier, A., Sarrazin, A., Thoison, O., DuBow, M., Faure, D. (2012). Efficient biostimulation of native and introduced quorum-quenching *Rhodococcus erythropolis* populations is revealed by a combination of analytical chemistry, microbiology, and pyrosequencing. *Appl. Environ. Microbiol.*, 78(2), 481–492.

Compant, S., Duffy, B., Nowak, J., Cleìment, C., Barka, E.A. (2005). Use of plant growth-promoting bacteria for biocontrol of plant diseases: Principles, mechanisms of action, and future prospects. *Appl. Environ. Microbiol.*, 71, 4951–4959.

Defoirdt, T. (2018). Quorum-sensing systems as targets for antivirulence therapy. *Trends Microbiol.*, 26(4), 313–328.

Delalande, L., Faure, D., Raffoux, A., Uroz, S., D'Angelo-Picard, C., Elasri, M., Carlier, A., Berruyer, R., Petit, A., Williams, P. et al. (2005). N-hexanoyl-L-homoserine lactone, a mediator of bacterial quorum-sensing regulation, exhibits plant-dependent stability and may be inactivated by germinating *Lotus corniculatus* seedlings. *FEMS Microbiol. Ecol.*, 52(1), 13–20.

Dessaux, Y. and Faure, D. (2018). Quorum-sensing and quorum quenching in *Agrobacterium*: A "go/no go system"? *Genes (Basel)*, 9(4), 210.

Dessaux, Y., Grandclément, C., Faure, D. (2016). Engineering the rhizosphere. *Trends Plant Sci.*, 21, 266–278.

Dong, Y.H., Wang, L.H., Xu, J.L., Zhang, H.B., Zhang, X.F., Zhang, L.H. (2001). Quenching quorum-sensing-dependent bacterial infection by an N-acyl homoserine lactonase. *Nature*, 411(6839), 813–817.

Dong, Y.H., Zhang, X.F., Xu, J.L., Zhang, L.H. (2004). Insecticidal *Bacillus thuringiensis* silences *Erwinia carotovora* virulence by a new form of microbial antagonism, signal interference. *Appl. Environ. Microbiol.*, 70(2), 954–960.

El Sahili, A., Kwasiborski, A., Mothe, N., Velours, C., Legrand, P., Moréra, S., Faure, D. (2015). Natural guided genome engineering reveals transcriptional regulators controlling quorum-sensing signal degradation. *PLoS One*, 10(11), e0141718.

Elias, M., Dupuy, J., Merone, L., Mandrich, L., Porzio, E., Moniot, S., Rochu, D., Lecomte, C., Rossi, M., Masson, P. et al. (2008). Structural basis for natural lactonase and promiscuous phosphotriesterase activities. *J. Mol. Biol.*, 379(5), 1017–1028.

Fuqua, W.C., Winans, S.C., Greenberg, E.P. (1994). Quorum sensing in bacteria – The LuxR-LuxI family of cell density-response transcriptional regulators. *J. Bacteriol.*, 176, 269–275.

Grandclément, C., Tannières, M., Moréra, S., Dessaux, Y., Faure, D. (2016). Quorum quenching: Role in nature and applied developments. *FEMS Microbiol. Rev.*, 40, 86–116.

Guendouze, A., Plener, L., Bzdrenga, J., Jacquet, P., Rémy, B., Elias, M., Lavigne, J.P., Daudé, D., Chabrière, E. (2017). Effect of quorum quenching lactonase in clinical isolates of *Pseudomonas aeruginosa* and comparison with quorum sensing inhibitors. *Front. Microbiol.*, 8, 227.

He, Y.W. and Zhang, L.H. (2008). Quorum sensing and virulence regulation in *Xanthomonas campestris*. *FEMS Microbiol. Rev.*, 32, 842–857.

Joshi, J.R., Khazanov, N., Khadka, N., Charkowski, A.O., Burdman, S., Carmi, N., Yedidia, I., Senderowitz, H. (2020). Direct binding of salicylic acid to *Pectobacterium* N-acyl-homoserine lactone synthase. *ACS Chem. Biol.*, 15(7), 1883–1891.

Kang, J.E., Han, J.W., Jeon, B.J., Kim, B.S. (2016). Efficacies of quorum sensing inhibitors, piericidin A and glucopiericidin A, produced by *Streptomyces xanthocidicus* KPP01532 for the control of potato soft rot caused by *Erwinia carotovora* subsp. *atroseptica*. *Microbiol. Res.*, 184, 32–41.

Kanugala, S., Kumar, C.G., Rachamalla, H.K.R., Palakeeti, B., Kallaganti, V.S.R., Nimmu, N.V., Cheemalamarri, C., Patel, H.K., Thipparapu, G. (2019). Chumacin-1 and Chumacin-2 from *Pseudomonas aeruginosa* strain CGK-KS-1 as novel quorum sensing signaling inhibitors for biocontrol of bacterial blight of rice. *Microbiol. Res.*, 228, 126301.

Kim, S.-R., Lee, K.-B., Kim, J.-E., Won, Y.-J., Yeon, K.-M., Lee, C.-H., Lim, D-J. (2015). Macroencapsulation of quorum quenching bacteria by polymeric membrane layer and its application to MBR for biofouling control. *J. Membrane Sci.*, 473, 109–117.

Koch, B., Liljefors, T., Persson, T., Nielsen, J., Kjelleberg, S., Givskov, M. (2005). The LuxR receptor: The sites of interaction with quorum-sensing signals and inhibitors. *Microbiology*, 151(11), 3589–3602.

Kuyukina, M.S., Ivshina, I.B., Kamenskikh, T.N., Bulicheva, M.V., Stukova, G.I. (2013). Survival of cryogel-immobilized *Rhodococcus* strains in crude oil-contaminated soil and their impact on biodegradation efficiency. *Int. Biodeterior. Biodegradation*, 84, 118–125.

Kwasiborski, A., Mondy, S., Chong, T.M., Chan, K.G., Beury-Cirou, A., Faure, D. (2015a). Core genome and plasmidome of the quorum-quenching bacterium *Rhodococcus erythropolis*. *Genetica*, 143(2), 253–261.

Kwasiborski, A., Mondy, S., Chong, T.M., Barbey, C., Chan, K.G., Beury-Cirou, A., Latour, X., Faure, D. (2015b). Transcriptome of the quorum-sensing signal-degrading *Rhodococcus erythropolis* responds differentially to virulent and avirulent *Pectobacterium atrosepticum*. *Heredity*, 114(5), 476–484.

Lang, J. and Faure, D. (2014). Functions and regulation of quorum-sensing in *Agrobacterium tumefaciens*. *Front. Plant Sci.*, 5, 14.

Latour, X., Delorme, S., Mirleau, P., Lemanceau, P. (2009). Identification of traits implicated in the rhizosphere competence of fluorescent pseudomonads: Description of a strategy based on population and model strain studies. In *Sustainable Agriculture*, Lichtfouse, E., Navarrete, M., Debaeke, P., Veìronique, S., Alberola, C. (eds). Springer, Amsterdam.

Latour, X., Barbey, C., Chane, A., Groboillot, A., Burini, J.F. (2013). *Rhodococcus erythropolis* and its γ-lactone catabolic pathway: An unusual biocontrol system that disrupts pathogen quorum sensing communication. *Agronomy*, 3, 816–838.

Manefield, M., Welch, M., Givskov, M., Salmond, G.P., Kjelleberg, S. (2001). Halogenated furanones from the red alga, *Delisea pulchra*, inhibit carbapenem antibiotic synthesis and exoenzyme virulence factor production in the phytopathogen Erwinia carotovora. *FEMS Microbiol. Lett.*, 205(1), 131–138.

Molina, L., Constantinescu, F., Michel, L., Reimmann, C., Duffy, B., Défago, G. (2003). Degradation of pathogen quorum-sensing molecules by soil bacteria: A preventive and curative biological control mechanism. *FEMS Microbiol. Ecol.*, 45(1), 71–81.

Paza, C., Cárcamo, G., Silva, M., Becerra, J., Urrutia, H., Sossa, K. (2013). Drimendiol, a drimane sesquiterpene with quorum sensing inhibition activity. *Nat. Prod. Commun.*, 8(2), 147–148.

Pun, M., Khazanov, N., Galsurker, O., Weitman, M., Kerem, Z., Senderowitz, H., Yedidia, I. (2021). Phloretin, an apple phytoalexin, affects the virulence and fitness of *Pectobacterium brasiliense* by interfering with quorum-sensing. *Front. Plant Sci.*, 12, 671807.

Raoul des Essarts, Y., Sabbah, M., Comte, A., Soulère, L., Queneau, Y., Dessaux, Y., Hélias, V., Faure D. (2013). N,N'-alkylated Imidazolium-derivatives act as quorum-sensing inhibitors targeting the *Pectobacterium atrosepticum*-induced symptoms on potato tubers. *Int. J. Mol. Sci.*, 14(10), 19976–19986.

Rodríguez, M., Torres, M., Blanco, L., Béjar, V., Sampedro, I., Llamas, I. (2020). Plant growth-promoting activity and quorum quenching-mediated biocontrol of bacterial phytopathogens by *Pseudomonas segetis* strain P6. *Sci. Rep.*, 10(1), 4121.

Ryan, P.R., Dessaux, Y., Thomashow, L.S., Weller, D.M. (2009). Rhizosphere engineering and management for sustainable agriculture. *Plant Soil*, 321, 363–383.

Shepherd, R.W. and Lindow, S.E. (2009). Two dissimilar N-acyl-homoserine lactone acylases of *Pseudomonas syringae* influence colony and biofilm morphology. *Appl. Environ. Microbiol.*, 75(1), 45–53.

Shinohara, M., Nakajima, N., Uehara, Y. (2007). Purification and characterization of a novel esterase (beta-hydroxypalmitate methyl ester hydrolase) and prevention of the expression of virulence by *Ralstonia solanacearum*. *J. Appl. Microbiol.*, 103(1), 152–162.

Uroz, S., Chhabra, S.R., Cámara, M., Williams, P., Oger, P., Dessaux, Y. (2005). N-Acylhomoserine lactone quorum-sensing molecules are modified and degraded by *Rhodococcus erythropolis* W2 by both amidolytic and novel oxidoreductase activities. *Microbiology*, 151(10), 3313–3322.

Uroz, S., Oger, P.M., Chapelle, E., Adeline, M.T., Faure, D., Dessaux, Y. (2008). A *Rhodococcus qsdA*-encoded enzyme defines a novel class of large-spectrum quorum-quenching lactonases. *Appl. Environ. Microbiol.*, 74(5), 1357–1366.

Vancov, T., Jury, K., Rice, N., Van Zwieten, L., Morris, S. (2006). Enhancing cell survival of atrazine degrading *Rhodococcus erythropolis* NI86/21 cells encapsulated in alginate beads. *J. Appl. Microbiol.*, 102, 212–220.

Weller, D.M. (2007). *Pseudomonas* biocontrol agents of soilborne pathogens: Looking back over 30 years. *Phytopathology*, 97, 250–256.

Yoshihara, A., Shimatani, M., Sakata, M., Takemura, C., Senuma, W., Hikichi, Y., Kai, K. (2020). Quorum sensing inhibition attenuates the virulence of the plant pathogen *Ralstonia solanacearum* species complex. *ACS Chem. Biol.*, 15(11), 3050–3059.

Yu, H., Liang, H., Qu, F., He, J., Xu, G., Hu, H., Li, G. (2016). Biofouling control by biostimulation of quorum-quenching bacteria in a membrane bioreactor for wastewater treatment. *Biotechnol. Bioeng.*, 113(12), 2624–2632.

Yu, H., Du, C., Qu, F., He, J., Rong, H. (2021). Efficient biostimulants for bacterial quorum quenching to control fouling in MBR. *Chemosphere*, 286(2), 131689.

Zhang, Y., Kong, J., Huang, F., Xie, Y., Guo, Y., Cheng, Y., Qian, H., Yao, W. (2018). Hexanal as a QS inhibitor of extracellular enzyme activity of *Erwinia carotovora* and *Pseudomonas fluorescens* and its application in vegetables. *Food Chem.*, 255, 1–7.

Zhang, W., Fan, X., Li, J., Ye, T., Mishra, S., Zhang, L., Chen, S. (2021). Exploration of the quorum-quenching mechanism in *Pseudomonas nitroreducens* W-7 and its potential to attenuate the virulence of *Dickeya zeae* EC1. *Front. Microbiol.*, 12, 694161.

8

Phage-mediated Biocontrol Against Plant Pathogenic Bacteria

Fernando CLAVIJO-COPPENS[1], Clara TORRES-BARCELO[2], Mireille ANSALDI[3], Nicolas TAVEAU[1] and Denis COSTECHAREYRE[1]

[1] *GREENPHAGE SAS, France*

[2] *INRAE-PACA, Unité de Pathologie Végétale, Avignon, France*

[3] *Laboratoire de Chimie Bactérienne, UMR 7283, CNRS Aix-Marseille University, France*

Plant pathogenic bacteria can cause severe damaging diseases every year, ranging from spots, mosaic patterns or pustules on leaves and fruits, or smelly tuber rots to plant death, and collectively they cause losses of over $1 billion dollars worldwide to the food production chain (Mansfield et al. 2012; Kannan et al. 2015). Antibiotics have been used against certain bacterial diseases with mixed results. Some antibiotics are absorbed by the plant and are distributed systemically. Most antibiotics are non-specific, acting not only against the target pathogen but also against other bacteria naturally present in the environment or plant and animal microflora (Svircev et al. 2018). Imprudent and massive use of antimicrobials in agriculture facilitated the emergence of antibiotic-resistant pathogens (Balogh et al. 2003; Manyi-Loh et al. 2018). However, the use of antibiotics on crops and ornamentals plants is prohibited

Biocontrol of Plant Disease,
coordinated by Claire PRIGENT-COMBARET and Bernard DUMAS. © ISTE Ltd 2022.

in the EU or highly regulated in the United States (Taylor and Reeder 2020). The use of chemicals to control bacterial infection on plants gives satisfactory control of some diseases when environmental conditions favor the development and spread of the pathogen. Unfortunately, the application of copper compounds and other agrochemicals has a strong environmental impact, and is often related to the appearance of bacterial resistance/tolerance or as problematic, a viable but non-culturable state (Um et al. 2013; Behlau et al. 2020; Kan et al. 2020).

To face dependence on the use of antibiotics and agrochemical products in agriculture, alternatives based on natural products are possible. Bacteriophages, also known informally as phages, have been used to treat infectious diseases since their discovery at the beginning of the 20th century (d'Hérelle 1917; Davison 1922). Phage therapy is a term restricted to human and animal applications, whereas, when phages are used for plant protection purposes, the term phage-based biocontrol is more adequate (Buttimer et al. 2017; Holtappels et al. 2021). Since the beginning of the 2000s, a growing demand for more ecological bactericides has reactivated the interest of the use of phages in the crop production industry, proposed as alternatives to the control of bacterial populations increasingly resistant to antibiotics and other agrochemicals in a very demanding field of antimicrobial products.

8.1. Introduction

8.1.1. *A tale of bacteriophages*

As is often the case in the history of science, the discovery of bacterial viruses (i.e. bacteriophages) took place in several phases in the wake of the early beginnings of virology. At that time, by the end of the 19th century, a virus was defined as a cell killing agent, invisible under a microscope and able to go through a Chamberland filter, which is a porous porcelain filter that retains particles with a size larger than 300 nm. The British bacteriologist H. Hankin (1865–1939) is the first author to have demonstrated the existence of a bactericidal agent in water. While working on the cholera bacillus in a government laboratory in India, he described in an eminent publication in 1896 the rapid destruction of the bacillus when a culture of it was put in contact with water from the Ganges and the Jumna

(Hankin 1896). He demonstrated that this agent passed through a Chamberland filter and was destroyed by autoclaving. This discovery was concomitant with rabies and tobacco mosaic virus ones, and they all experimentally reveal the existence of new infectious agents that attack human cells in the case of rabies and plants in the case of the tobacco mosaic virus or bacteria and whose size is smaller than that of bacteria because they pass through filters that retain the latter (Ansaldi et al. 2020a).

Frederick Twort (1877–1950) was the second bacteriologist to deal directly with invisible antibacterial agents. He confirmed their ultra-filterable nature and their inactivation by heat and also showed that these particles no longer grow when the sample is free of bacteria as described in his 1915 publication (Twort 1915). Two years later, in 1917, the French–Canadian researcher Félix d'Hérelle (1873–1949), while working at the Pasteur Institute, laid down certain fundamentals of virology: he described the isolation and specificity of these "antagonistic" agents as well as the formation of lysis patches on a bacterial lawn (d'Herelle 1917). He proposed the name "bacteriophage", from the ancient Greek bakterion (small stick) and phágos (eater), and was the first to use them therapeutically in the context of the fight against bacillary dysentery caused by *Shigella*, thus giving rise to phagotherapy (D'Herelle 1918). His collaborator Georges Eliava returned to Tbilisi, the capital of Georgia, and founded the therapeutic and research center that is still performing phage therapy nowadays. Phage therapy is now considered as a safe and versatile alternative to antibiotics, especially when antibiotic use is either ineffective in the case of multiresistant bacteria or prohibited as in crop production (Melo et al. 2020; Holtappels et al. 2021).

Phage research is also deeply associated with the birth and development of molecular biology. The contributions of the study of phages to modern molecular biology are very important and used on a daily basis without sometimes remembering the origin of these techniques, from in vitro transcription systems to site-specific recombination cloning systems (Ansaldi et al. 2020a). Other examples are spectacular applications of anti-phage systems, better known as molecular scissors or CRISPR-Cas9 that allow genome editions of most living organisms from unicellular to highly organized multicellular organisms (Doudna and Charpentier 2014). This particular application recently led Jennifer Doudna and Emmanuelle Charpentier to receive the Nobel Prize in Chemistry in 2020.

The first use of phages against a phytopathogenic bacterium probably took place in 1924 with *Xanthomonas campestris* pv. *campestris* as a model. W. Mallman and C. Hemstreet found that the filtrate of the liquid collected from decaying cabbage inhibited the growth of the bacteria that cause black rot in cabbage (Mallmann and Hemstreet 1924). The first field trials were carried out against the blight of Stewart's corn caused by the plant pathogen *Pantoea stewartii*. Maize seeds infected with *P. stewartii* were soaked with a phage preparation isolated from dead maize plants, infected seeds and infected plants in the recovery phase. Plants grown from seeds treated with phage filtrate showed a significant reduction in disease compared to untreated control plots (Thomas 1935). However, this type of research was overlooked because the nature of the phage was poorly understood at that time and data on its effectiveness was limited.

In recent years, a number of phage biocontrol applications in crop production have been developed and assayed at different scales including in laboratory, greenhouses and field conditions (see section 8.2.1). What is crucial for a successful biocontrol application is to characterize the phages to be applied in depth at a genomic level as well as to regard their interactions with the pathogen to treat to avoid failures due to resistance acquisition or inappropriate host range (both too narrow or too large host ranges could be problematic).

8.1.2. *Phage biology, cycle, host range and resistance mechanisms*

Most phages in public databases are tailed phages (Caudoviridae) containing double-stranded DNA (kingdom Duplodnaviridae); however, a variety of morphological and genomic structures (DNA, RNA, single-stranded, double-stranded) can be found. Compared to bacterial genomes and those of giant viruses, phage genomes are relatively small, ranging from 15 to a few hundred kb; however, these genomes exhibit high genetic diversity and complex phylogenetic evolution, particularly due to horizontal gene transfer events and their propensity to recombine (Dion et al. 2020).

Phages can essentially propagate in two ways: the lytic cycle or the lysogenic pathway. A virulent phage is capable of performing only a lytic cycle that results in host lysis upon release of the new virion progeny. The lytic cycle starts as a classic viral cycle with the virus binding to a specific receptor on the host surface. Phages then inject their genetic material, allowing the synthesis of the first proteins that come to action, involved in the hijacking of the host functions and counteraction of the host defenses. The genome is replicated according to its genetic nature, generally using the host's machinery. Then, structural as well as proteins involved in virion assembly are synthesized. With the notable exception of filamentous phages that are budding from the bacterial membrane, the lytic cycle ends up by the production and the timely assembly of lysis proteins that make the cell burst.

On the other hand, a temperate phage can either perform a lytic cycle or enter lysogeny, during which its genome usually integrates within the bacterial chromosome, and is then called a prophage. In case a stress is detected by the host, the prophage is induced and manages to hijack the stress signaling pathways and then starts a productive lytic cycle in order to infect new host cells (Calendar 2006).

The host recognition stage is crucial and decisive for the outcome of the infection and varies according to the type of host considered (Gram-positive or Gram-negative, presence of a capsule or not, extracellular appendages or not, etc.). Each bacteriophage recognizes one or more components exposed on the host surface in a very specific way. The specificity of this interaction constrains the host spectrum of the bacteriophages to bacteria that expose a given receptor. The interaction occurs via bacteriophage receptor binding proteins (RBPs) located at the end of the caudal fibers, and the nature of the receptors varies according to the characteristics of the bacterial envelope (Bertozzi Silva et al. 2016). Unlike antibiotics, phages generally show a narrow host range, specific for a single bacterial species. Often phages display a large host range, able to infect only a few strains of the same species (Gill and Abedon, 2003). Only few exceptions exist, with phages displaying a broader host range by infecting taxonomic trans-order strains (Kaiser and Dworkin 1975; Jensen et al. 1998; Peters et al. 2015).

In phage therapy, as well as for biocontrol applications, we should always ask what are the characteristics of a good phage candidate? It is important to have a deep knowledge of the phages used, such as their life cycle and their virulence characteristics towards the host. Indeed, temperate phages present a priori disadvantages for therapeutic approaches because, during lysogenic conversion, they can confer on the host properties making it either more virulent or more resistant to infection by a new phage or undetectable by the immune system (Monteiro et al. 2019). In addition, bacteria encode various phage resistance mechanisms such as restriction–modification (RM) systems, involving restriction enzymes that degrade foreign genetic elements, or the now very famous CRISPR-cas systems directed against foreign genetic elements. Host-phage coevolution allowed the emergence of anti-resistance mechanism such as Anti-CRISPR (Acr) inhibitor proteins or anti-RM systems (Tock and Dryden 2005; Maxwell 2017). These systems can provide interesting properties in the context of a therapeutic usage of phages and should be looked at when possible.

8.1.3. *Phage ecology on plant-associated ecosystem*

Phages are a natural evolutionary force with wide consequences for their bacterial hosts and ecosystems in general (Schwartz and Lindell 2017; Koskella and Taylor 2018). The effects of phages on bacteria are diverse, with quantitative effects dramatically decreasing bacterial populations and qualitative effects such as selecting resistant bacterial variants. Phages are part of the plant-associated microbiota, but very little is known about their prevalence in agricultural environments and role in the ecology and emergence of plant bacterial diseases, compared to other domains (Koskella and Taylor 2018). Numerous studies reveal the importance of phages for human health and disease, and that most phages in the human gut are prophages, revealing very specific interactions with important consequences: secretion of phage-encoded toxins, modulation of the bacterial envelope, mediation of bacterial infectivity and the control of bacterial gene regulation (Mirzaei and Maurice 2017; Schroven et al. 2021). Metagenomic analyses in marine aquatic environments have revealed that phages play key roles in the density and diversity of microbial communities (Schwartz and Lindell 2017). In plant eukaryotic viruses, metagenomic approaches have revealed the association of their distribution with agriculture land use (Bernardo et al. 2018). A recent metavirome study suggests that phages dynamically challenge the bacterial community in Antarctic soils, as revealed by the correlation with local specific CRISPR-arrays and the prevalence of the phage exclusion (BREX) and Defense Island System Associated with Restriction–Modification (DISARM) bacterial defense systems (Bezuidt et al. 2020). Complex microbial communities in soil, water and plants could be better understood by studying phages, perhaps decisive factors for plant pathogenic bacteria consequences on plant disease.

Plant-associated environments are particularly challenging environments to study phages because of the intricate spatial structure and abiotic stresses present. In the phyllosphere, phages face inhospitable conditions, severely inactivated by the direct UV light and desiccation, suggesting that lysogenic cycles may be preferred by temperate phages (Gill and Abedon 2003). Withal, long-term studies demonstrated a high prevalence and local adaptation of phages on leaves of horse chestnut trees (Koskella et al. 2011). Phages are particularly abundant in the rhizosphere, as demonstrated by metagenomics and microscopic visualization, with high phage:bacteria ratios (Takahashi et al. 2011). In these habitats, phages could be protected from UV light and humidity is preserved, but they could be inactivated by conditions such as extreme pH or soil absorption, depending on the specific physico-chemical traits of phages and matrices (Ye et al. 2019). Interestingly, heavy metals, such as copper, used to combat plant pathogens, have diverse effects on both hosts and parasites. For instance, copper possesses an antimicrobial activity selecting for resistant variants, as well as induces the excision of prophages from bacterial genomes (Lee et al. 2006; Seiler and Berendonk 2012).

Several inoculation assays demonstrate that phages translocate and remain inside plants for days and months, as in melon or vineyards, but it is not known if they naturally penetrate in a recurrent way and its long-term effects (Das et al. 2015; Rahimi-Midani and Choi 2020). Only few studies suggest the presence and transmission of phages by insect vectors, responsible for spreading diseases such as Pierce's disease or fire blight; they have been searched for and sometimes found at very low concentration (Woods et al. 1981; Bhowmick et al. 2016; Clavijo-Coppens et al. 2021). Phages are likely to be present in irrigation water of agricultural settings, as suggested by several studies (Kavagutti et al. 2019; Cazares et al. 2021; Trotereau et al. 2021). Phage abundance in all those environments may depend on the seasons and most likely correlate with prevalence of their bacterial hosts (Gill and Abedon 2003). Phage transfer between plant-associated niches is probably related to the movement of the bacterial hosts, including wind and rain, animals and human activities (Gill and Abedon 2003). A deeper exploration of these questions could reveal the role of phages in the evolution and pathogenicity of the bacterial pathogens.

8.2. Bacteriophages for plant health

Plant pathogenic bacteria infect a wide range of hosts in most regions on Earth and in almost all types of ecosystems. Among pathogenic bacteria, the most represented species are *Erwinia* spp, *Pectobacterium* spp, *Pantoea* spp, *Pseudomonas* spp, *Ralstonia* spp, *Burkholderia* spp, *Acidovorax* spp, *Dickeya* spp, *Xanthomonas* spp, *Clavibacter* spp and *Xylella* spp (Mansfield et al. 2012; Kannan et al. 2015). Today, the management of plant pathogenic bacteria relies on integrated approaches, including the adoption of Good Agricultural Practices to reduce the proliferation of outbreaks, but most frequently, the use of chemical treatments and antibiotic formulations. The proliferation of bacteria resistant to antibiotics and other bactericides, as well as the potentially adverse effects of agrochemicals on the environment could have a detrimental effect on fighting bacterial diseases on crop production (Cooksey 1994; Pietrzak and McPhail 2004; Fair and Tor 2014). However, to control crop pathogens, regulatory restrictions, public health concern and development of resistant strains require new sustainable alternative control measures.

8.2.1. *Phage-based biocontrol against plants pathogens*

Currently, a number of phage products are being developed or are already commercially available for a variety of plant pathogens in vitro and in vivo (Buttimer et al. 2017; Svircev et al. 2018; Vu and Oh 2020; Ansaldi et al. 2020b). Suitable in vitro potential of phages does not necessarily correspond to successful biocontrol potential in planta. Interestingly, different phage-based biocontrol application strategies are described in different studies. Those studies described encouraging results for the biocontrol of agricultural production following infection strategies under realistic condition (Buttimer et al. 2017; Holtappels et al. 2021). Optimization of phage-based biocontrol requires a series of field and greenhouse experiments, since in vitro trials may not always fully reflect the harsh conditions of the real environment and of the different crops to cure.

In Table 8.1, we summarize a large number of pathosystem crop models that have been investigated over the last 20 years, with a specific emphasis towards studies performed in greenhouses and in open-field conditions. Most of these studies were performed on perennial crops with very promising results, while on annual plants, the efficacy of phage-based biocontrol depends on the crop as well as on the formulation. In addition, all these studies of phage biocontrol were assessed to control Gram-negative bacteria, with the exception of the Gram-positive plant pathogenic actinomycete *Clavibacter michiganensis*.

Bacterial target	Crop	Proof of concept	Phage formulation	Formulation association	Reducing pathogen incidence	References
Acidovorax citrulli	Watermelon	Greenhouse	Single phage	None	86%	Rahimi-Midani et al. (2020)
	Melon	Greenhouse	Single phage	None	66%	Rahimi-Midani and Choi, (2020)
Burkholderia glumae and *B. plantarii*	Rice	Greenhouse	Single phage	None	33–100%	Adachi et al. (2012)
Clavibacter michiganensis subsp. *nebraskensis*	Maize	Greenhouse	Single phage	None	seeds (58–76%), seedlings (20%)	Kimmelshue et al. (2019)
Dickeya solani	Potato	Field	Single phage	None	ND	Adriaenssens et al. (2012)
Pectobacterium atrosepticum and *P. carotavorum* subsp. *carotavorum*	Potato	Field	Cocktail 6 phages	None	95–97%	Zaczek-Moczydłowska et al. (2020b)
	Onion	Field	Cocktail 4 phages	None	75–85%	Zaczek-Moczydłowska et al. (2020a)
	Potato	Field	Single phage	None	90%	Voronina et al. (2019)
	Lettuce	Greenhouse	Single phage	None	80%	Lim et al. (2013)
Pseudomonas syringae pv. *actinidiae*	Kiwi	Greenhouse	Cocktail 4 phages	None	75%	Flores et al. (2020)
P. syringae pv. *porri*	Leek	Greenhouse	Single phage	None	ND decreased lesion length	Rombouts et al. (2016)
		Fields	Cocktail 6 phages	None	25%	
P. syringae pv. *syringae* and *P. syringae* pv. *morsprunorum*	Cherry trees	Greenhouse	Single phage	None	10–60%	Rabiey et al. (2020)
			Cocktail 7 phages	None	30–50%	
			Cocktail 13 phages	None	10–50%	

Bacterial target	Crop	Proof of concept	Phage formulation	Formulation association	Reducing pathogen incidence	References
P. tolaasii	Oyster Mushroom	Greenhouse	Single phage	None	100%	Kim et al. (2011)
Ralstonia solanacearum	Banana	Greenhouse	Cocktail 2 phages	None	100%	Ramírez et al. (2020)
	Potato	Garden pots	Cocktail 6 phages	None	20%	Wei et al. (2017)
	Tomato	Greenhouse	Single phage	None	100%	Bae et al. (2012)
		Greenhouse	Single phage	None	100%	Elhalag et al. (2018)
		Greenhouse	Single phage	None	~50–~90%	Álvarez et al. (2019)
			Cocktail 3 phages	None	100%	
		Greenhouse	Single phage	None	ND	Wang et al. (2019)
			Cocktail 4 phages	None	80%	
		Field	Single phage	None	ND	
			Cocktail 4 phages	None	80%	
Xanthomonas axonopodis pv. *allii*	Onion	Field	Cocktail Agriphage, OmniLytics	None	26%	Lang et al. (2007)
				Cu-Mz	26%	
				ASM	35%	
X. axonopodis pv. *citri*	Grapefruit	Greenhouse	Single phage	None	1,4%	Balogh et al. (2008)
				Cu-Mz	32%	
			Cocktail 5 phages	None	55%	
				SMP	12%	
X. axonopodis pv. *citrumelo*	Citrus		Cocktail 3 phages	SMP	48%	
				Cu-Mz	83%	

Bacterial target	Crop	Proof of concept	Phage formulation	Formulation association	Reducing pathogen incidence	References
X. campestris pv. *campestris*	Broccoli	Greenhouse	Single phage	Nonpathogenic Xanthomonas	41–86%	Nagai et al. (2017)
		Field	Single phage	Nonpathogenic Xanthomonas	39%	
X. campestris pv. *juglandis*	Walnut	Greenhouse	Cocktail 6 phages	None	not significantly different	McNeil et al. (2001)
X. campestris pv. *vesicatoria*	Tomato	Greenhouse	Cocktail Agriphage; OmniLytics	None	1%	Balogh et al. (2003)
				PCF	30%	
				Casecrete	51%	
				SMP	62%	
		Field	Cocktail Agriphage; OmniLytics	None	14%	
				PCF	24%	
				Casecrete	32%	
				SMP	18%	
		Greenhouse	Cocktail Agriphage; OmniLytics	None	96%	Flaherty et al. (2000)
		Field	Cocktail Agriphage; OmniLytics	None	17%	
		Field	Cocktail Agriphage; OmniLytics	None	28%	Obradovic et al. (2004)
				HP	25%	
				ASM	39%	
				CuMz	51%	
				HP + ASM	23%	

Bacterial target	Crop	Proof of concept	Phage formulation	Formulation association	Reducing pathogen incidence	References
X. citri pv. *citri*	Citrus	Greenhouse	Cocktail 6 phages	None	61%	Ibrahim et al. (2017)
				ASM	76%	
		Field	Cocktail 6 phages	ASM	50%	
				SMP + sucrose + ASM	84%	
	Grapefruit	Greenhouse	Single phage	None	20–60% and one phage shows more symptoms that control	Balogh et al. (2018)
X. euvesicatoria	Pepper	Greenhouse	Single phage	None	33–58%	Gašić et al. (2018)
				Cu-Mz	87%	
		Greenhouse	Cocktail 2 phages	SMP	21%	Tewfike and Shimaa (2015)
				PCF + sucrose	18,30%	
		Field	Cocktail 2 phages	SMP	24,50%	
				PCF + sucrose	32,20%	
X. euvesicatoria pv. perforans	Tomato	Greenhouse	Single phage	None	21–56%	Balogh et al. (2018)
X. oryzae pv. *oryzae*	Rice	Field	Cocktail 3 phages	SMP	51%	Chae et al. (2014)
			Cocktail 3 phages h-mutants	SMP	54%	
		Greenhouse	Single phage	SMP	31%	Dong et al. (2018)
Xylella fastidiosa subsp. *fastidiosa*	Grapevines	Greenhouse	Cocktail 4 phages	None	50%	Das et al. (2015)

Cu-Mz = Copper hydroxide Mancozeb; ASM = Acibenzolar-S-Methyl; SMP = Skim Milk Powder; PCF = Pregelatinized Corn Flour; HP = Harpin Protein

Table 8.1. *Proof of concept of phage biocontrol for crop protection during the last 20 years*

Acidovorax phages

Acidovorax citrulli is a seed-borne bacterial pathogen listed on the A1 list of pests recommended for regulation as quarantine pests by the European and Mediterranean Plant Protection Organization (EPPO A1 List 2020). This pathogenic bacterium is passively carried on seeds and contaminated soils, causing plant diseases in watermelon and melon production. To our knowledge, only two *A. citrulli* phages have been recently isolated (Rahimi-Midani et al. 2018, 2020). Phage-based biocontrol assay, under greenhouse conditions, suggests that a single application of ACPWH-phage reduces pathogen incidence caused by *A. citrulli* in melon and watermelons by soil drenching the phage solution. In addition, symptoms caused by *A. citrulli* did not progress on phage-treated plants, suggesting a phage translocation from soil to leaf tissue and an arrest of symptom progression (Rahimi-Midani and Choi 2020; Rahimi-Midani et al. 2020).

Burkholderia *and* Clavibacter *phages*

Burkholderia glumae and *B. plantarii* are seed-borne pathogens on rice, which are an increasingly important disease problem in global rice production (Zhou-qi et al. 2016; Akimoto-Tomiyama, 2021). Adachi and colleagues report that a prophylactic phage treatment on rice seeds, coating with single phage preparations, reduced the percentage of diseased seedlings on rice infected by *B. glumae* and *B. plantarii* (Adachi et al. 2012). Similar results were reported to the control by phages against *Clavibacter michiganensis* subsp. *nebraskensis* (*Cmn*), another seed-borne pathogen in maize and listed as a quarantine pest by the EPPO (EPPO A2 List 2020). By seed coating CN8-phage, *Cmm* incidence is reduced by 76, 58 and 20% on seeds, germinated seeds and seedling, respectively, bringing the question about CN8-phage phage translocation efficacy (Kimmelshue et al. 2019). These results suggest that phage coating is an interesting alternative to ensure disease-free seeds for commercialization, eliminating seed-borne bacteria and avoiding undesirable bacterial pathogen propagation of quarantine listed species.

Dickeya *phages*

Dickeya solani causes blackleg and soft rot in potato crops causing heavy losses in field, and also during storage and transit (Charkowski et al. 2020). The effect of phage treatment against *D. solani* on potato tuber and plant growth was examined in a field trial on potato. Adriaenssens and colleagues measured a small reduction in disease incidence and a higher yield at the end of the bioassay on potato tubers treated by a single-phage application (Adriaenssens et al. 2012). Many other *Dickeya* phages were isolated and tested in vitro to protect post-harvest potatoes (Czajkowski et al. 2011; Adriaenssens et al. 2012; Carstens et al. 2018). More bioassays in planta are needed to prove the efficacy of phage biocontrol against *D. solani* under crop conditions.

Pectobacterium *phages*

To control *Pectobacterium atrosepticum*, causing blackleg on potato under fields conditions, Zaczek-Moczydłowska and colleagues demonstrated that phage cocktail application reduces pathogen incidence by 75–97% on potato tubers under field conditions (Zaczek-Moczydłowska et al. 2020a, 2020b). In addition, the potential of phages specific to *P. carotovorum* subsp. *carotovorum* as biocontrol agents has been trialed both in laboratory and in field assays. Voronina and colleagues showed that single-PP16 phage application on potato plants reduce 90% of *P. carotovorum* incidence in a field trial (Voronina et al. 2019). Moreover, a single-phage PP1 application also reduced pathogen incidence at 80% under greenhouse conditions on lettuce plants infected by *P. carotovorum* subsp. *carotovorum* (Lim et al. 2013). These studies show that single-phage treatment can produce results as good as phage cocktails for the biocontrol of *Pectobacterium* spp. in planta.

Pseudomonas syringae *phages*

Pseudomonas syringae is a species complex of bacterial plant pathogens that infects over 180 plant species. Many globally important crops fall within the host range of *P. syringae*, making the pathogen one of the most economically destructive (Xin et al. 2018). This species is able to cause disease in a wide range of economically relevant crops such as tomato, kiwifruit, bean, leek and cherry trees (Rabiey et al. 2020). Rombouts and colleagues observed that a phage cocktail directed against *P. syringae* pv. *porri* reduced bacterial blight symptoms in leek. However, under greenhouse conditions, the authors suggested only a reduction of lesion length on leaf, while on a field trial, phage cocktail reduced pathogen incidence (Rombouts et al. 2016). The same tendency was confirmed in a greenhouse trial by priming leek seeds: phages KIL3b and KIL5 enhanced seed germination by 25% compared to leek blight-infected seeds (Holtappels et al. 2020). Phage biocontrol was also assayed against the quarantine plant pest *P. syringae* pv. *actinidiae* (EPPO A2 List 2020). Kiwifruit plants sprayed with CHF1, CHF7, CHF19 and CHF21 phage-cocktail showed a reduction of 75% of symptoms compared to untreated plants; cocktail application also reduced the bacterial canker incidence under greenhouse conditions (Flores et al. 2020).

For *P. syringae* pv. *syringae*, causing bacterial canker on cherry trees, phage efficacy was tested under greenhouse conditions. On micro-propagated cherry-infected plantlets, a reduction of pathogen incidence around 42% or 50% was observed when cocktails containing 5, 7 and 13 phages, respectively, were sprayed. All phage cocktails reduced bacterial population density almost immediately after application. In addition, results on infected plants sprayed by single phages reduce pathogen incidence from 10% to 80% (Rabiey et al. 2020). Likewise, *P. syringae pv. syringae* phage-based biocontrol under laboratory conditions are comparable to greenhouse results on the cherry tree model (Akbaba and Ozaktan, 2021).

P. syringae phages prevent pathogen propagation on seeds and micro-propagated plants. Also, phage-use manages plant pest on infected trees as kiwi fruits and potentially limits quarantine bacterial species dissemination.

Ralstonia *phages*

Ralstonia solanacearum species complex has an unusually wide host range and is also included in the A2 listed as a quarantine pest by the EPPO (EPPO A2 List 2020) infecting more than 200 species belonging to more than 50 botanical families, including economically important crops such as tomato, potato, tobacco, banana and eggplant (Hayward 1991; Denny 2006; Mansfield et al. 2012). Phages active on *R. solanacearum*, isolated from environmental sources, have been characterized under laboratory conditions (Wei et al. 2017; Addy et al. 2019; Wang et al. 2019; Trotereau et al. 2021). *R. solanacearum* and its host tomato plants are one of the most represented models of phage-biocontrol in planta. Single-phage φRSL1 application was successfully decreasing by 100% of the symptoms caused by *R. solanacearum* on tomato seedlings under growth chamber conditions (Fujiwara et al. 2011). Phage-based preparations have been tested also in single-phage applications of PE204, RsPod1EGY and vRsoP-WR2 by soil drenching or injection demonstrating a 100% reduction of pathogen incidence under greenhouse conditions (Bae et al. 2012; Elhalag et al. 2018; Álvarez et al. 2019). Similar results were obtained when tomato plants were treated with a phage-cocktail containing vRsoP-WF2, vRsoP-WM2 and vRsoP-WR2 (Álvarez et al. 2019). Compared to this result, single-phage preparations suggested by Álvarez and colleagues and also by Wang and colleagues exhibited less reducing effect, with a reduction of the pathogen incidence on tomato between 50% and 80%. These results suggest that while all phage types were effective at reducing the incidence of bacterial wilt disease on their own, they were also effective when applied in combinations of cocktails.

Under greenhouse conditions, *R. solanacearum* phages M5 and M8 isolated from banana-contaminated soil were reduced by 100% of bacterial wilt symptoms on contaminated plants as well as the pathogen incidence on contaminated soils (Ramírez et al. 2020). In addition, despite very good efficacy on tomato and banana plants, some phage combinations were reduced with less efficiency of the pathogen incidence, but were still efficient for other plant pathosystems. For example, in potato plants infected by *R. solanacearum*, a phage cocktail applied by injection only attenuates 20% of the pathogen incidence (Wei et al. 2017). Wei and colleagues also demonstrated that phage cocktail application reduced pathogenic bacteria from contaminated soils.

Erwinia amylovora *phages*

Erwinia amylovora is a plant pathogenic bacterium, listed on the EPPO A2 list as a quarantine pest (EPPO A2 List 2020). *E. amylovora* causes the well-known fire blight disease of apple and pear trees (Vanneste 2000). *Erwinia* spp. forms biofilms, an important virulence factor, which is involved in the resistance of the phytopathogens to some bactericides as well as protects cells from phage infection (Koczan et al. 2009; Born et al. 2017). Potential of phage biocontrol has been trialed under laboratory conditions on apple and pear tissues, suggesting a reduction of pathogen incidence after treatment by a phage cocktail (Schwarczinger et al. 2017). Spraying on pear and apple tissues of a phage formulation combining a single phage with a carrier bacteria also reduced pathogen incidence (Boulé et al. 2011). Under laboratory conditions, *E. amylovora* phages exhibited good therapeutical potential (Ravensdale et al. 2007; Thompson et al. 2019). Their potential use as biocontrol agents needs more trials both in the laboratory and in field. In addition, under laboratory conditions, the genetically modified *Erwinia* phage Y2::dpoL1-C enhances biofilm degradation (Born et al. 2017).

Xylella *phages*

Listed as a quarantine pest (EPPO A2 List 2020), *Xylella fastidiosa* is a Gram-negative xylem-specialized bacterium, emerging in Asia and Europe and associated with numerous socio-economically important plant diseases (Denancé et al. 2017; Rapicavoli et al. 2018). In spite of the slow growing and fastidious characteristics to manipulate this plant pathogen under laboratory conditions, a few phages active on *X. fastidiosa* have been isolated. In vitro assays suggested that isolated phages manifested a lytic activity with a narrow host range among *X. fastidiosa* strains isolated from the Americas and in Europe (Ahern et al. 2014; Clavijo-Coppens et al. 2021). In the absence of efficient and authorized control methods to eliminate this bacterium in planta, Das and colleagues presented the first step forward to phage-biocontrol against *X. fastidiosa* in planta. Under greenhouses conditions, a phage cocktail composed of Sano, Salvo, Prado and Paz phages, and injected directly in infected grapevine plants, reduced the pathogen levels and prevented *X. fastidiosa* symptom development (Das et al. 2015). This study is an encouraging starting point to develop phage biocontrol against *Xylella* spp.; however, more isolated phages are needed to propose phages formulation against all *Xylella* subspecies and to develop application methods.

Xanthomonas *phages*

Xanthomonas species are all clearly distinctive in their pathology and host targets. *Xanthomonas* infections occur on at least 124 monocotyledonous and 268 dicotyledonous plant spp. (Leyns et al. 1984). Several phage-based biocontrol have been used to manage several plant diseases caused by *Xanthomonas* spp. (Stefani et al. 2021). Of note, on perennial plants, plant pathogens may be more difficult to control by phage applications than on annual crops.

On the one hand, to illustrate this proposal on perennial plants, Saccardi and colleagues demonstrate that peach leaves infected by *X. campestris* pv. *pruni* sprayed with single F8-phage displayed less symptoms than untreated leaves (Saccardi et al. 1993). Meanwhile, in the face of a large number of studies with positive results previously shown on other pathosystems, the efficacy of phage biocontrol on in planta presents some limitations. In greenhouses, phage cocktail treatment against *X. campestris* pv. *juglandis* did not show any effect to protect walnuts trees against infection (McNeil et al. 2001). Efficacy of phage-based biocontrol was tested on grapefruit plants against *X. citri* pv. *citri*, listed as a quarantine pest (EPPO A1 List 2020), and *X. axonopodis*, pv. *citri* under greenhouse and nursery conditions. Single-phage application of φXaacA1 had no effect on *X. axonopodis*, pv. *citri*, and the integration of phage and copper-mancozeb resulted in equal or less pathogen control than copper-mancozeb application alone (Balogh et al. 2008, 2018). In addition, φXv3-21 has deleterious effects, increasing about 25% of symptoms caused by *X. citri* pv. *citri* on grapefruits trees (Balogh et al.

2008). At the same time, Balogh and colleagues demonstrated a positive effect of phage treatment, controlling *X. citri* pv. *citri* by assembling four phages in a cocktail. This cocktail application reduced the grapefruit citrus canker incidence (Balogh et al. 2008). The efficacy of a phage cocktail was subsequently improved by adding plant Systemic Acquired Resistance (SAR) inducers and/or phage-protective formulations to the phage preparations (see section 1.2.2.2). Under field conditions, pathogen incidence on citrus plants infected by *X. citri* pv. *citri* was reduced by 50% using SAR inducers and by 84% using SAR inducers and phage-protective formulations (Ibrahim et al. 2017). Under greenhouse conditions, Balogh and colleagues obtained a good efficacy on citrus plants infected by *X. axonopodis* pv. *citrumelo* by improving phage persistence on leaves sprayed with a three-phage cocktail supplemented only by protective formulations. Furthermore, phage cocktail supplemented by copper-mancozeb reduces 83% of bacterial spots caused by *X. axonopodis* pv. *citrumelo* compared to untreated plants (Balogh et al. 2008). Taken together, all those results suggest that phage-formulation needs to be adapted to each pathosystem to improve phage efficacy, in particular on trees.

On the other hand, to control *Xanthomonas* spp on annual crops, few examples demonstrated a promising efficacy under greenhouses or field conditions. The commercial phage preparation Agriphage from OmniLytix, supplemented by phage-protective formulations, reduced onion leaf blight and tomato leaf spot severity caused by *X. axonopodis* pv. *allii* and *X. campestris* pv. *vesicatoria*, respectively. Phage-based preparation had previously been observed to be more effective than copper treatment in suppressing and reducing the disease severity (Flaherty et al. 2000; Balogh et al. 2003; Obradovic et al. 2004; Lang et al. 2007). Furthermore, the commercial phage mixture that AgriPhage developed from diverse host bacterial strains improved disease suppression. In addition, phage application on infected plants enhanced weight gain of extra-large fruit on tomato plants as compared to copper bactericides (Flaherty et al. 2000; Balogh et al. 2018). For *X. oryzae* Phage-biocontrol, a large panel of phages active on *X. oryzae* pv. *oryzae* has been isolated (Chae et al. 2014). Chae and colleagues demonstrated that formulated phage cocktails reduced the pathogen occurrence on mature rice plants. All these results suggest that phage preparations use on rice is a good alternative to control pathogenic bacteria from seed to developed plants (Adachi et al. 2012; Chae et al. 2014). More phages should be isolated and tested using similar methods in order to facilitate the use of phage cocktails in field experiments.

8.2.2. *Optimizing a phage-based preparation*

The major challenges associated with control plant pathogenic bacteria stem from the fact that plant production is carried out in an extremely hostile environment, presenting biotic and abiotic factors that may impair phage persistence. We highlighted phages as a promising tool and successfully evaluated to control economically important plant pathogens. In order to maximize phage effectiveness on crop productions, phage quality and stability, phage mixtures, association with protective compounds and/or amendment with other microorganisms, the method of application and timing, should all be optimized.

8.2.2.1. *Phage-related information to consider*

Most of well-characterized phages deployed on plant pathogenic biocontrol are members of the phage order *Caudoviridae*. Tailed phages have dsDNA genomes and complex protein capsids. In terms of life cycle, tailed phages may be either virulent or temperate (Ackermann 1998). Strictly virulent phages are preferred for therapy, thereby avoiding problems caused by lysogeny (Howard-Varona et al. 2017; Fillol-Salom et al. 2019). Besides the life cycle, phage genomic information could also be necessary to anticipate undesirable phage-coded genes such toxins, transposases or integrases (Krüger and Lucchesi 2015; Yoshikawa et al. 2018).

For crop protection purposes, phages are deposited on generally hydrophobic surfaces where they are subjected to a number of adverse elements such as sunlight irradiation, cycles of desiccation and rehydration, fluctuating ambient temperatures and pH (Buttimer et al. 2017). Ideally, selected phages for biocontrol should be stable for those characteristics under lab conditions in order to anticipate field conditions (Czajkowski et al. 2014; Pinheiro et al. 2019).

Given their inability to persist for long periods under extreme and hostile environments, phage preparations require being applied at high titers (Iriarte et al. 2012; Balogh et al. 2018). The highest the titer of phages, the greatest their ability to inhibit the growth of the bacterial host, and in less time (Álvarez et al. 2019; Pinheiro et al. 2019; Zaczek-Moczydłowska et al. 2020a; Akbaba and Ozaktan 2021).

Efficacy of phages in vitro in reducing bacterial growth is not always reflected in vivo and realistic conditions should be considered when selecting a phage candidate for biocontrol (Henry et al. 2013). In some cases, formulations with only a single phage reduce incidence of bacterial plant disease (Kim et al. 2011; Adachi et al. 2012; Rahimi-Midani et al. 2020). Single-phage use might be ineffective because bacteria rapidly develop resistance under a single selective pressure exerted (Fujiwara et al. 2011; Wang et al. 2019). To prevent resistance emergence, the use of phage cocktails targeting different host receptors, instead of individual strains, is likely to prevent resistance development (Wei et al. 2017; Wang et al. 2019).

Currently, there are many guidelines for selecting a phage to be used in a phage cocktail, but the most important requirement is to make sure that the phage cocktail can multiply and disseminate in the environment, and specifically lyse the targeted bacteria (Buttimer et al. 2017).

8.2.2.2. *Phage-protective formulations*

Abiotic factors dramatically constrain the stability of phages and are probably the Achilles heel of releasing phages in agriculture habitats. Yet this weakness, together with their extreme specificity, probably makes them the safest biocontrol agents. Few studies have investigated the mid-term viability of phages in plant-associated environments, showing that at least some phages remain viable for weeks in plants and soil (Das et al. 2015; Rahimi-Midani and Choi 2020). Further experiments are required to assess phage viability in a variety of environmental conditions, but it is presumed to be short, and the phages applied on crops have never been found to reach the human food chain.

As phages are easily propagated in bacteria cultivated in liquid media, these formulations are the most popular phage preparations. Moreover, liquid formulations prevent phages from drying and inactivation. One of the greatest challenges in using bacteriophages for plant disease control is their extremely short residual activity in the phyllosphere. Several studies indicated that phage populations can drop to undetected levels hours after applications (Balogh, 2002; Iriarte et al. 2007). Direct sunlight significantly reduces phage persistence, and the efficacy of phages is greatly reduced due to its short longevity on plant leaf surfaces (Czajkowski et al. 2011; Born et al. 2015). Additionally, formulations are required to address the negative environmental conditions such as strong UV light, effect of desiccation as well as limited phage persistence on the plant surface, maintaining high phage concentrations. In order to enhance phage vitality and longevity on foliage, several studies have sought to apply protective formulations such as skimmed milk powder, casecrete, sucrose and corn flour. Those additives increase phage populations longevity in greenhouse and open-field trials (Balogh et al. 2003, 2008; Ibrahim et al. 2017).

Phage associations with other microorganisms, playing a role of surrogated host, increased phage abundance by supporting phage replication and protecting phages to be inactivated while bacterial host is absent (Jones et al. 2007; Balogh et al. 2010; Boulé et al. 2011; Álvarez et al. 2019). In addition, carrier bacteria should be able to colonize plants, playing a role of antagonists to the target bacterium.

8.2.2.3. *Phage application associating other biocontrol methods*

Antibacterial compounds

Plant pathogenic bacteria control currently relies on multiple applications of agro-chemicals or antibiotics-based bactericides when they are allowed. In this regard, it is also essential to analyze the potential interactions between phages and other agrochemicals commonly used in agriculture, such as copper-based compounds. Copper and mancozeb are among the most widely used bactericides and fungicides in agricultural productions. The combination of phage and copper-mancozeb treatments did not result in increased pathogen control over copper-mancozeb alone. On the contrary, plants treated with this combination were as severely diseased as the plants treated with phage only, suggesting that copper probably negatively impacted phage survival if applied in a tight spray schedule (Balogh et al. 2008). Conversely, phage combination with copper-hydroxide and acibenzolar-S-methyl was the most effective combination in fields, reducing the disease severity by 96–98% (Šević et al. 2019).

Recently, Abrahamian and colleagues reported that combination of phages, cymoxanil, famoxadone and phosphoric acid, significantly improved disease management compared to the copper-based standard treatment (Abrahamian et al. 2019). More studies on the effects of copper and other heavy metals on phages are needed to establish their inhibitory or other effects on phages (Lee et al. 2006).

SAR inducer compounds

Alternative strategies have been tested and include activation of natural plant defense mechanisms by systemic acquired resistance (SAR) inducers (Vallad and Goodman 2004). Mixtures of phages and the plant defense activators, such acibenzolar-S-methyl (ASM) or the harpin protein elicitors, were evaluated under field and greenhouse conditions. Those approaches provided disease control equal to, or greater than, either treatment alone, reducing the disease severity (Balogh et al. 2003; Obradovic et al. 2004; Lang et al. 2007). ASM and harpin protein were used in combination with copper oxychloride or copper hydroxide; they did not effectively reduce pathogen incidence on foliage under field conditions compared with copper alone (Obradovic et al. 2004; Ibrahim et al. 2017).

Others microorganisms

New methods explored involve applying phage along with attenuated bacterial strains which served both as propagating hosts for the phage and as antagonists of the target bacterium (Iriarte et al. 2012). Combined application of phages with others bacterial biocontrol agents, such as nonpathogenic strains to antagonize pathogenic strains, or plant growth-promoting rhizobacteria (PGPR) that induce SAR, have been successfully used in several pathosystems (Obradovic et al. 2004).

8.2.2.4. *Application strategy*

The application of phages generally provides a significant reduction in bacterial spot severity compared with the untreated control. The stabilizing solutions mentioned above may have provided phage protection from environmental factors. Bacteria often occupy protected spaces, for example, inside cankers, under buds, or in the soil matrix, where they could be unavailable to phages, and phage-delivery strategies should be designed. Phage-based preparations are easy to apply on crop productions by: (i) seed-coating or seed priming to eradicate seed-borne pathogens; (ii) soil drenching against soilborne pathogens and (iii) spraying aerial plant parts to control foliar diseases.

In agriculture, antibacterial seed coatings can control bacterial plant diseases by protecting the seeds from contaminated plant residue, contaminated soil and contamination within the seeds. Bacteriophage-polymer coatings, which are stable during drying and storage, and are compatible with ecological systems, provide an alternative to traditional antibiotics in agriculture. Applying phages directly to the soil is a viable and likely the most unobtrusive, scalable method of application. Preparation of applied phages in the field through irrigation water may represent an innovative environmentally friendly method for a safe and sustainable agriculture (Iriarte et al. 2012). In addition, phages are most likely to be effective against foliar pathogens because of the direct access to these pathogens. Less accessible pathogenic bacteria such as woody cankers, xylem-restricted or against a soilborne vascular pathogen in a field need higher volumes of phages to cover and saturate the infected plants, and phage-based control is less efficient (Balogh et al. 2010; Das et al. 2015).

A deep knowledge of the ecology and epidemiology of the bacteria and environment where phages will be applied is necessary in order to use the best adapted phages in the appropriate place and at the correct moment. This often implies narrow windows of phage treatments and the need to optimize the application regimes. Prophylactic application demonstrated a good phage efficacy if treatment was applied one hour, one day or even three weeks before inoculation with the pathogen (Civerolo and Keil 1969; Das et al. 2015). Applying phages to plants in the evening or in the morning could be effective for maintaining high phage concentrations on the leaf surface for an extended period of time and was associated with improved disease control by reducing phage inactivation by sun UV radiation (Flaherty et al. 2000; Balogh et al. 2003). Application intervals depend on the pathosystem, and less spaced timing and frequency of application control with higher-efficacy pathogen incidence (Flaherty et al. 2000; Lang et al. 2007; Balogh et al. 2010).

8.2.2.5. *Lifetime of phages applied for crop protection*

Phages applied in the agricultural environment face complex biotic and abiotic conditions. Weather conditions, soil types and plant varieties are some of the factors influencing microbes in agricultural environments. Land use, water supply and quality, global warming, agricultural practices and microbiota composition are other important variables that can have an impact on the fate of phages and their bacterial hosts. Will the applied phages succeed in controlling pathogens and favor crops or will they be inactivated or not enough efficient? Several factors will determine the positive or neutral outcome of phages in the environment. To date, no negative effects of their application have been identified.

On the biotic side of the agricultural environment, the applied phages will have a significant effect on the targeted bacterial pathogens but not on other bacteria, as proven by recent microbiota analysis (Wang et al. 2019). Besides their effect of reducing bacterial population densities, the fact that bacteria can develop resistance to phages was considered as negative in the past. However, as with many bacteria, their ability to resist to phages has been repeatedly linked to costs on important pathogen life-history traits, which ultimately decreased aggressiveness (degree of damage to the host)

(Wendling et al. 2017). For example, phage-resistant *R. solanacearum* bacteria had a reduced growth rate, carrying capacity and competitive ability compared to non-resistant strains (Wang et al. 2019). Defective lipopolysaccharide (LPS) mutants of the same bacterium, as a result of exposure to phage, resulted in no wilting symptoms in tobacco plants, likely due to evolutionary trade-offs (Hendrick and Sequeira 1984). In plant pathogens such as *E. amylovora*, *P. syringae* or *X. fastidiosa* and *Xanthomonas* spp, bacterial modifications of phage attachment molecules involve virulent factors such as LPS, flagella and pili (Evans et al. 2010; Ahern et al. 2014; Holtappels et al. 2020). In few cases, phage resistance in plant pathogenic bacteria did not result in a reduction of virulence. Under laboratory conditions, phage-resistant bacteria could have the opposite effect on increasing pathogen virulence (Yamada et al. 2007; Addy et al. 2012).

Experimental evidence by Britt Koskella and colleagues proved that phage resistance has costs for bacteria plant colonization, and that the probability of resistance evolution is impaired in plants compared to in vitro conditions for *P. syringae* (Meaden et al. 2015; Hernandez and Koskella 2019). Studies on *P. protegens* (former *P. fluorescens*) and Rhizobia demonstrate local adaptation of phages to these host bacteria, with phages infecting less efficiently distant bacterial communities (Koskella et al. 2011; Van Cauwenberghe et al. 2021). All this data suggests that phage resistance in bacteria is harder to evolve in vivo than in vitro and that, in the event that it happens, the consequences of plant disease may benefit plant health. Phages may not be the magic bullet to erase plant disease completely, but including them in an integrated pest management may increase the chances for a safer and balanced microbiome.

8.2.3. *Pros and cons of phage biocontrol on crop protection*

There are currently a number of challenges in the implementation of phage therapy in agriculture; the pros and cons are summarized in Table 8.2.

Phages are self-replicating and natural entities, as well as easy to isolate from various sources such as soil, water and plants. In open fields, they are quickly degraded in the absence of the target host (Wang et al. 2019). Phage control generally requires the identification of the target pathogen. Unlike broad-spectrum chemical antibiotics or bactericidal compounds, phages are highly host-specific, with no impact on non-target organisms nor eukaryotic cells and reducing dysbiosis risk on plants (Frampton et al. 2014; Flores et al. 2020; Rabiey et al. 2020). This narrow specificity may hinder the value of phages in agriculture especially in situations in which more than one closely related pathogens are simultaneously present.

Pros	Cons
– Fairly easy and inexpensive to produce and application can be carried out with standard farm equipment; – specific or highly discriminatory, eliminating only target bacteria; – non-toxic for soil, water, plants, animals and humans; – self-replicating and self-limiting as long the targeted bacteria is present in the environment; – efficacy improved by co-application with many agrochemicals and other bacterial control methods; – adaptable solution against multi-resistant bacteria or other bacteria species; – biofilm degradation.	– Phage characterization for only use strictly virulent; – quickly degraded in the absence of the host; – phage inactivation under environmental conditions (UV, dessiccation, etc.); – single-phage application favors phage-resistant bacteria development; phage-cocktail could minimize this risk; – narrow host range.

Table 8.2. *Phage biocontrol against plant pathogens: pros and cons*

In addition, since phages consist mostly of nucleic acids and proteinaceous shells, phage-based products represent low environmental impact, limiting the release of persistent/dangerous chemicals into the environment. Strictly virulent phages can effectively and specifically lyse their bacterial targets increasing the viral particles number. As phages need bacteria with an active metabolism in order to successfully infect other target cells, their activity stops once the target has been destroyed (Frampton et al. 2014; Abedon et al. 2017).

The use of phage as biological control agents requires knowledge of the host specificity and an understanding of the way in which the host bacterium and phage interact, tailored towards a specific pathogen (Buttimer et al. 2017; Pirnay et al. 2018). All the individual phages in a cocktail should be genomically characterized to ensure that no known virulence and antibiotic-resistant genes are included in their genomes. Current standards of phage production demand a deep phage characterization evading undesirable phage-coding genes such as toxin and virulence genes or genes related to lysogeny. While bacteria are susceptible to phages, they also have many mechanisms to protect themselves from phages, antibiotics and environmental stress. *Erwinia* spp, *Xanthomonas* spp and *Xylella* spp form biofilms, an important virulence factor that is involved in the resistance of phytopathogens to certain bactericides (Rodrigues et al. 2008; Koczan et al. 2009; Yan et al. 2012). Improvement of biofilm degradation using phages enhances EPS-degrading properties bypassing biofilm protection to phages (Born et al. 2017; Yoshikawa et al. 2018; Holtappels et al. 2020).

Phage-resistant bacteria emerge when single-phage preparations or narrow host range phage strains are applied (Frampton et al. 2014). Phage-resistant bacteria primarily result from the mutation of phage-triggered receptors, and some of them are essential for bacteria pathogenesis. This reduces pathogenic fitness impacting the maximum growth rate of pathogen, carrying capacity and competitive ability against non-resistant ancestral strains (Wang et al. 2019). Jackson and colleagues suggested that phages evolved in turn. Spontaneously derived from their wild-type parent phages, trained phages lyse not only parent wild-type bacteria but also bacteriophage-resistant mutants evolved from parental population (Jackson 1989; Flaherty et al. 2000). Moreover, the application of bacteriophages in a cocktail should minimize the emergence risk of resistant bacteria. In addition, a clear advantage is that phage-based control can be adapted to multi-resistant pathogens or any bacterial species such as quarantine pathogens, and when chemical control is not allowed owing to legal regulations (Rahimi-Midani et al. 2020; Clavijo-Coppens et al. 2021).

Although phages may be effective in one specific crop, they may not be as effective in other related pathosystems (Gill and Abedon 2003; Iriarte et al. 2012). The production of phages predominantly involves a combination of host growth and subsequent purification. Phage preparations are fairly easy and relatively not expensive to produce and can be stored at 4°C for months without significant titer reduction (Rombouts et al. 2016; Gašić et al. 2018). In most cases, applications can be carried out with standard farm equipment with versatile applications for liquid formulations and may lead to the development of phage combinations in which individual members work in synergy to eliminate the target bacteria (Born et al. 2011). For vascular plant diseases, phage applications could need specific material to procure the phage preparation and reach the targeted pathogenic bacteria (Das et al. 2015).

8.3. Phage-based biocontrol regulations

Biocontrol methods use living organisms or natural substances to prevent or reduce damage caused by harmful organisms such animal pests, weeds and microbial pathogens. Four categories of biocontrol agents are established: (i) macroorganisms such as insects and nematodes; (ii) microorganisms such as viruses, bacteria or fungi; (iii) chemical mediators such as pheromones and (iv) natural substances from mineral, plant or animal origin. Biocontrol is a tool that can be used in the context of crop protection and its principle is based on managing the balance of populations of aggressors rather than on their total eradication. The use of bacteriophages fits perfectly into this eco-friendly approach because of their nature and biological action with a specific lytic activity against the targeted bacteria and its transient lifetime.

Interestingly, phages in agriculture in the United States and Canada benefit from the status of generally recognized as safe (GRAS), which greatly simplifies their regulation by the US Food and Drug Administration (FDA). The use of phage preparations for food safety purposes has also received approval in several other countries, such as Israel, Canada, Switzerland, Australia and New Zealand. These approvals are frequently issued based upon US regulatory approvals (Vikram et al. 2020). In contrast, the European Food Safety Authority (EFSA) is slow to change and adapt regulations that delay the possibility of phage biocontrol in the European Union. In the UE, phages belong to the regulatory category of Microbial Plant Protection Products (PPP), which includes bacteria, fungi and viruses. Before registration and marketing authorization, PPPs must undergo a thorough risk assessment to food and environmental safety assays. These PPPs are subject to the Regulation (EC) No. 1107/2009. The most widely used authorized microorganism in the world is the entomophagous bacterium *Bacillus thuringiensis*, which is authorized in organic farming as a biocontrol agent against pine and oak processionary caterpillars.

For phage-based biocontrol products in Europe, the regulatory context is the same as that for conventional crop protection products. Moreover, biocontrol products that can be used in organic farming are at the crossroads of three regulations at French and European levels and must respect all these requirements:

– the substance must be approved under the EC regulation no. 1107/2009;

– the commercial phage-based preparation must have a marketing authorization;

– the substance must be authorized by the EC regulation no. 889/2008, which lists in annex II all the substances of crop protection products that can be used in organic farming.

The EC regulation no. 1107/2009 process for a formulation of phages as a PPP takes several years, and the total costs may reach 2–5 million euros, with variable fees depending on the member states (Marchand 2017). Indeed, the application procedure requires the constitution of a complete evaluation dossier of the finalized product, which is evaluated by the EFSA and includes physical and chemical analyses, efficacy evaluation, toxicity and ecotoxicology. As of today, no bacteriophage cocktail has been authorized as PPPs in EU, which does not provide an effective biocontrol solution for plant pathogenic bacteria. A major problem with phage-mediated biocontrol is that bacteriophage cocktails need to be updated regularly in order to be efficient against new emerging bacterial strains and any change to one of the components of a PPP requires reregistering. Encouragingly, the FDA and USDA have started to allow such updates of some commercial phage preparations without the need to go through the entire regulatory approval process with every update (Vikram et al. 2020).

However, in agriculture, preparations of bacteriophages against phytopathogenic bacteria represented pioneering biocontrol agents (Mallmann and Hemstreet 1924). Only few phage-based products for plant biocontrol have reached the market. The first company to receive legal authorization to commercialize a phage-based preparation was Omnilytics, Inc. (https://www.omnilytics.com/). This US-based company proposes four different biopesticide containing phages: (i) for more than 15 years, this company produces and sells AgriPhage® to control bacterial spot in tomatoes and peppers caused by *X. campestris* pv. *vesicatoria*, or *P. syringae* pv. *tomato* for conventional and organic farming; (ii) Agriphage CMM® to control bacterial canker in tomatoes caused by *C. michiganensis* subsp. *michiganensis*; (iii) AgriPhage-Citrus® canker against *X. citri* subsp. *Citri* and (iv) AgriPhage-Fire Blight® to control the fire blight caused by *E. amylovora*, one of the most problematic diseases for apple and pear trees in Europe. Enviroinvest Zrt. company based in Hungary developed Erwiphage Plus® for the control of fire blight in apple trees (https://www.erwiphage.com/).

The Scottish company APS Biocontrol Ltd. developed Biolyse®, a phage-based mixture used for the post-harvest control on potatoes tubers to prevent soft rot disease caused by *Pectobacterium* spp. during storage (https://www.apsbiocontrol.com/). It should be noted that the bacteriophage preparations of APS Biocontrol have been approved as "processing aids", a regulatory category without the need of authorization under EC regulation no. 1107/2009, unlike products classified as PPP for field use (Ansaldi et al. 2020b). More recently, the company Otsuka Pharmaceutica Co., Ltd. proposed the phage-based solution XylPhi-PD™, officially approved by the US Environmental Protection Agency (EPA), available for commercial sale in the United States and commercialized by A&P Inphatec, LLC to reduce the symptoms of Pierce's Disease in grapevines caused by *X. fastidiosa* subsp. *fastidiosa* (https://inphatec.com/).

In the UE, legislation governing phage biocontrol may need to become more malleable for the best application and performance of phage products as biopesticides. Thus, evolution of UE regulation and national initiatives, such as the French Ministry of Agriculture's EcoPhyto 2 plan, could encourage the development of new phage-based biocontrol products.

8.4. Conclusions and perspectives

In the crop industry, many pathogens have become resistant to antibiotics and some agrochemicals probably due to their abuse and misuse (Balogh et al. 2003; Manyi-Loh et al. 2018). Since the early 2000s, a growing demand for bio-bactericides has revived the interest in the use of bacteriophages that are therefore proposed as alternatives for the control of bacterial populations in agricultural model (Table 8.1) with a high demand for antimicrobial products. However, there is a need for more research to improve the effectiveness of phage biocontrol and more solutions to the technical challenges.

Every phyllosphere and plant-associated environment is unique and complex; therefore, a deep knowledge of the ecology and epidemiology of the targeted bacteria and the environment where phages will be applied is necessary. Many mistakes were made during early trials of phage-based biocontrol, and most can be attributed to a poor understanding of the biological nature and application of phages (Gill and Hyman 2010; Lin et al. 2017; Vandamme and Mortelmans 2019). Abiotic factors dramatically constrain the stability of phages and are probably the Achilles heel of releasing phages in agriculture habitats. Yet this weakness, together with their extreme specificity, probably makes them the safest biocontrol agents. Few studies have investigated the mid-term viability of phages in plant-associated environments, showing that at least some phages remain viable for weeks in plants and soil (Das et al. 2015; Rahimi-Midani and Choi 2020). Further experiments are required to assess this, but it is presumed to be short, and phages applied in crops have never found to reach the human food chain. Protective formulations may give phages viability facing harsh environmental conditions such as strong UV light (Born et al. 2015). More studies on the effects of copper and other heavy metals on phages are needed to establish their inhibitory or other effects (Lee et al. 2006). The use of carrier non-pathogenic bacteria is a common strategy to enhance both viability and concentration of phages released to the environment (Álvarez et al. 2019). Water-associated applications, prophylaxis and non-pathogenic carrier bacteria are the most preferred strategies to improve phage efficacy and persistence in the environment (Iriarte et al. 2012). Nowadays, phage-based efficacy against plant pathogenic bacteria has been demonstrated at many levels, to control seed-borne and soilborne pathogens or to control mostly foliar diseases (Buttimer et al. 2017; Holtappels et al. 2021; Kimmelshue et al. 2019; Rahimi-Midani et al. 2020). Phage biocontrol also offers an alternative solution for fastidious plant diseases where conventional treatments are not available to treat multi-resistant bacteria (Das et al. 2015; Zhao et al. 2019). In particular, phage-based biocontrol must be included in the integrating pest management systems to reduce the need for agrochemicals. Phage biocontrol is still difficult to implement in open fields due to the poor development of formulations to effectively treat hectares of plants crops and/or in greenhouse conditions and the modern farming practices that rely on chemical pesticides that may be deleterious to the phages.

Finally, perception of phages by the public remains a black box. Phages have not gained widespread acceptance as compared to commercially proven pharmaceutical antimicrobial agents. Despite the approval to use phages in food and medical industries by several international agencies, FDA legislation governing phage biocontrol may need to become more malleable in the EU for better applications and performances of phage products as biopesticides.

8.5. Acknowledgments

We thank Dr. René BALLY for his valuable comments and his help in the preparation of the manuscript.

8.6. References

Abedon, S.T., García, P., Mullany, P., Aminov, R. (2017). Editorial: Phage therapy: Past, present and future. *Front. Microbiol.*, 8, 981.

Abrahamian, P., Jones, J.B., Vallad, G.E. (2019). Efficacy of copper and copper alternatives for management of bacterial spot on tomato under transplant and field production. *Crop Prot.*, 126, 104919.

Ackermann, H.W. (1998). Tailed bacteriophages: The order caudovirales. *Adv. Virus Res.*, 51, 135–201.

Adachi, N., Tsukamoto, S., Inoue, Y., Azegami, K. (2012). Control of bacterial seedling rot and seedling blight of rice by bacteriophage. *Plant Dis.*, 96, 1033–1036.

Addy, H.S., Askora, A., Kawasaki, T., Fujie, M., Yamada, T. (2012). The filamentous phage ϕRSS1 enhances virulence of phytopathogenic *Ralstonia solanacearum* on tomato. *Phytopathology*, 102, 244–251.

Addy, H.S., Ahmad, A.A., Huang, Q. (2019). Molecular and biological characterization of Ralstonia phage RsoM1USA, a new species of P2virus, isolated in the United States. *Front. Microbiol.*, 10, 267.

Adriaenssens, E.M., Vaerenbergh, J.V., Vandenheuvel, D., Dunon, V., Ceyssens, P.-J., Proft, M.D., Kropinski, A.M., Noben, J.-P., Maes, M., Lavigne, R. (2012). T4-related bacteriophage limestone isolates for the control of soft rot on potato caused by "*Dickeya solani*". *PLoS One*, 7.

Ahern, S.J., Das, M., Bhowmick, T.S., Young, R., Gonzalez, C.F. (2014). Characterization of novel virulent broad-host-range phages of *Xylella fastidiosa* and *Xanthomonas*. *J. Bacteriol.*, 196, 459–471.

Akbaba, M. and Ozaktan, H. (2021). Evaluation of bacteriophages in the biocontrol of *Pseudomonas syringae* pv. *syringae* isolated from cankers on sweet cherry (*Prunus avium* L.) in Turkey. *Egypt. J. Biol. Pest Control*, 31, 35.

Akimoto-Tomiyama, C. (2021). Multiple endogenous seed-born bacteria recovered rice growth disruption caused by *Burkholderia glumae*. *Sci. Rep.*, 11, 4177.

Álvarez, B., López, M.M., Biosca, E.G. (2019). Biocontrol of the major plant pathogen *Ralstonia solanacearum* in irrigation water and host plants by novel waterborne lytic bacteriophages. *Front. Microbiol.*, 10, 2813.

Ansaldi, M., Boulanger, P., Brives, C., Debarbieux, L., Dufour, N., Froissart, R., Gandon, S., Le Hénaff, C., Petit, M.-A., Rocha, E., Torres-Barceló, C. (2020a). A century of research on bacteriophages. *Virol. Montrouge Fr.*, 24, 9–22.

Ansaldi, M., Boulanger, P., Brives, C., Debarbieux, L., Dufour, N., Froissart, R., Gandon, S., Le Hénaff, C., Petit, M.-A., Rocha, E., Torres-Barceló, C. (2020b). Les applications antibactériennes des bactériophages. *Virol. Montrouge Fr.*, 24, 23–36.

Bae, J.Y., Jing, W., Hyoung, J.L., Eun, J.J., Senthilkumar, M., Eunsook, C., Seon-Woo, L. (2012). Biocontrol potential of a lytic bacteriophage PE204 against bacterial wilt of tomato. *J. Microbiol. Biotechnol.*, 22, 1613–1620.

Balogh, B. (2002). Strategies of improving the efficacy of bacteriophages for controlling bacterial spot of tomato. Thesis, University of Florida, Gainesville.

Balogh, B., Jones, J.B., Momol, M.T., Olson, S.M., Obradovic, A., King, P., Jackson, L.E. (2003). Improved efficacy of newly formulated bacteriophages for management of bacterial spot on tomato. *Plant Dis.*, 87, 949–954.

Balogh, B., Canteros, B.I., Stall, R.E., Jones, J.B. (2008). Control of citrus canker and citrus bacterial spot with bacteriophages. *Plant Dis.*, 92, 1048–1052.

Balogh, B., Jones, J., Iriarte, F., Momol, M. (2010). Phage therapy for plant disease control. *Curr. Pharm. Biotechnol.*, 11, 48–57.

Balogh, B., Nga, N.T.T., Jones, J.B. (2018). Relative level of bacteriophage multiplication in vitro or in phyllosphere may not predict in planta efficacy for controlling bacterial leaf spot on tomato caused by *Xanthomonas perforans*. *Front. Microbiol.*, 9, 2176.

Behlau, F., Gochez, A.M., Jones, J.B. (2020). Diversity and copper resistance of *Xanthomonas* affecting citrus. *Trop. Plant Pathol.*, 45, 200–212.

Bernardo, P., Charles-Dominique, T., Barakat, M., Ortet, P., Fernandez, E., Filloux, D., Hartnady, P., Rebelo, T.A., Cousins, S.R., Mesleard, F. et al. (2018). Geometagenomics illuminates the impact of agriculture on the distribution and prevalence of plant viruses at the ecosystem scale. *ISME J.*, 12, 173–184.

Bertozzi Silva, J., Storms, Z., Sauvageau, D. (2016). Host receptors for bacteriophage adsorption. *FEMS Microbiol. Lett.*, 363(4).

Bezuidt, O.K.I., Lebre, P.H., Pierneef, R., León-Sobrino, C., Adriaenssens, E.M., Cowan, D.A., Van de Peer, Y., Makhalanyane, T.P. (2020). Phages actively challenge niche communities in antarctic soils. *mSystems*, 5, e00234-20.

Bhowmick, T.S., Das, M., Heinz, K.M., Krauter, P.C., Gonzalez, C.F. (2016). Transmission of phage by glassy-winged sharpshooters, a vector of *Xylella fastidiosa*. *Bacteriophage*, 6, e1218411.

Born, Y., Fieseler, L., Marazzi, J., Lurz, R., Duffy, B., Loessner, M.J. (2011). Novel virulent and broad-host-range *Erwinia amylovora* bacteriophages reveal a high degree of mosaicism and a relationship to enterobacteriaceae phages. *Appl. Environ. Microbiol.*, 77, 5945–5954.

Born, Y., Bosshard, L., Duffy, B., Loessner, M.J., Fieseler, L. (2015). Protection of *Erwinia amylovora* bacteriophage Y2 from UV-induced damage by natural compounds. *Bacteriophage*, 5, e1074330.

Born, Y., Fieseler, L., Thöny, V., Leimer, N., Duffy, B., Loessner, M.J. (2017). Engineering of Bacteriophages Y2:dpoL1-C and Y2:luxAB for efficient control and rapid detection of the fire blight pathogen, *Erwinia amylovora*. *Appl. Environ. Microbiol.*, 83(12).

Boulé, J., Sholberg, P.L., Lehman, S.M., O'Gorman, D.T., Svircev, A.M. (2011). Isolation and characterization of eight bacteriophages infecting *Erwinia amylovora* and their potential as biological control agents in British Columbia, Canada. *Can. J. Plant Pathol.*, 33, 308–317.

Buttimer, C., McAuliffe, O., Ross, R.P., Hill, C., O'Mahony, J., Coffey, A. (2017). Bacteriophages and bacterial plant diseases. *Front. Microbiol.*, 8, 34.

Calendar, R. (ed.) (2006). *The Bacteriophages*, 2nd edition. Oxford University Press, Oxford.

Carstens, A., Djurhuus, A., Kot, W., Jacobs-Sera, D., Hatfull, G., Hansen, L., (2018). Unlocking the potential of 46 new bacteriophages for biocontrol of *Dickeya Solani*. *Viruses*, 10, 621.

Cazares, D., Cazares, A., Figueroa, W., Guarneros, G., Edwards, R.A., Vinuesa, P. (2021). A novel group of promiscuous podophages infecting diverse gammaproteobacteria from river communities exhibits dynamic intergenus host adaptation. *mSystems*, 6, e00773-20.

Chae, J.-C., Hung, N.B., Yu, S.-M., Lee, H.K., Lee, Y.H. (2014). Diversity of bacteriophages infecting *Xanthomonas oryzae* pv. *oryzae* in paddy fields and its potential to control bacterial leaf blight of rice. *J. Microbiol. Biotechnol.*, 24, 740–747.

Charkowski, A., Sharma, K., Parker, M.L., Secor, G.A., Elphinstone, J. (2020). Bacterial diseases of potato. In *The Potato Crop: Its Agricultural, Nutritional and Social Contribution to Humankind*, Campos, H. and Ortiz, O. (eds). Springer International Publishing, Cham.

Civerolo, E.L. and Keil, H.L. (1969). Inhibition of bacterial spot of peach foliage by *Xanthomonas pruni* bacteriophage. *Phytopathology*, 12, 1966–1967.

Clavijo-Coppens, F., Ginet, N., Cesbron, S., Briand, M., Jacques, M.-A., Ansaldi, M. (2021). Novel virulent bacteriophages infecting Mediterranean isolates of the plant pest *Xylella fastidiosa* and *Xanthomonas albilineans*. *Viruses*, 13, 725.

Cooksey, D.A. (1994). Molecular mechanisms of copper resistance and accumulation in bacteria. *FEMS Microbiol. Rev.*, 14, 381–386.

Czajkowski, R., Pérombelon, M.C.M., van Veen, J.A., van der Wolf, J.M. (2011). Control of blackleg and tuber soft rot of potato caused by *Pectobacterium* and *Dickeya* species: A review. *Plant Pathol.*, 60, 999–1013.

Czajkowski, R., Ozymko, Z., Lojkowska, E. (2014). Isolation and characterization of novel soilborne lytic bacteriophages infecting *Dickeya* spp. biovar 3 ('*D. solani*'). *Plant Pathol.*, 63, 758–772.

Das, M., Bhowmick, T.S., Ahern, S.J., Young, R., Gonzalez, C.F. (2015). Control of Pierce's disease by phage. *PLoS One*, 10, e0128902.

Davison, W.C. (1922). The bacteriolysant therapy of bacillary dysentery in children: Therapeutic application of bacteriolysants; d'Hérelle's pheromon. *Am. J. Dis. Child.*, 23, 531–534.

Denancé, N., Legendre, B., Briand, M., Olivier, V., de Boisseson, C., Poliakoff, F., Jacques, M.-A. (2017). Several subspecies and sequence types are associated with the emergence of *Xylella fastidiosa* in natural settings in France. *Plant Pathol.*, 66, 1054–1064.

Denny, T. (2006). Plant pathogenic *Ralstonia* species. In *Plant-Associated Bacteria*, Gnanamanickam, S.S. (ed.). Springer Netherlands, Dordrecht.

Dion, M.B., Oechslin, F., Moineau, S. (2020). Phage diversity, genomics and phylogeny. *Nat. Rev. Microbiol.*, 18, 125–138.

Dong, Z., Xing, S., Liu, J., Tang, X., Ruan, L., Sun, M., Tong, Y., Peng, D. (2018). Isolation and characterization of a novel phage Xoo-sp2 that infects *Xanthomonas oryzae* pv. *oryzae*. *J. Gen. Virol.*, 99, 1453–1462.

Doudna, J.A. and Charpentier, E. (2014). Genome editing. The new frontier of genome engineering with CRISPR-Cas9. *Science*, 346, 1258096.

Elhalag, K., Nasr-Eldin, M., Hussien, A., Ahmad, A. (2018). Potential use of soilborne lytic Podoviridae phage as a biocontrol agent against *Ralstonia solanacearum*. *J. Basic Microbiol.*, 58, 658–669.

EPPO A1 List (2020). Pests absent from the EPPO region [Online]. Available at: https://www.eppo.int/ACTIVITIES/plant_quarantine/https%3A%2F%2Fwww.eppo.int%2FACTIVITIES%2Fplant_quarantine%2FA1_list [Accessed 19 July 2021].

EPPO A2 List (2020). Pests locally present in the EPPO region [Online]. Available at: https://www.eppo.int/ACTIVITIES/plant_quarantine/https%3A%2F%2Fwww.eppo.int%2FACTIVITIES%2Fplant_quarantine%2FA2_list [Accessed 19 July 2021].

Evans, T.J., Trauner, A., Komitopoulou, E., Salmond, G.P.C. (2010). Exploitation of a new flagellatropic phage of *Erwinia* for positive selection of bacterial mutants attenuated in plant virulence: Towards phage therapy. *J. Appl. Microbiol.*, 108, 676–685.

Fair, R.J. and Tor, Y. (2014). Antibiotics and bacterial resistance in the 21st century. Perspect. *Med. Chem.*, 6, 25–64.

Fillol-Salom, A., Alsaadi, A., de Sousa, J.A.M., Zhong, L., Foster, K.R., Rocha, E.P.C., Penadés, J.R., Ingmer, H., Haaber, J. (2019). Bacteriophages benefit from generalized transduction. *PLoS Pathog.*, 15, e1007888.

Flaherty, J.E., Somodi, G.C., Jones, J.B., Harbaugh, B.K., Jackson, L.E. (2000). Control of bacterial spot on tomato in the greenhouse and field with h-mutant bacteriophages. *HortScience*, 35, 882–884.

Flores, O., Retamales, J., Núñez, M., León, M., Salinas, P., Besoain, X., Yañez, C., Bastías, R. (2020). Characterization of bacteriophages against *Pseudomonas syringae* pv. actinidiae with potential use as natural antimicrobials in kiwifruit plants. *Microorganisms*, 8(7), 974.

Frampton, R.A., Taylor, C., Holguín Moreno, A.V., Visnovsky, S.B., Petty, N.K., Pitman, A.R., Fineran, P.C. (2014). Identification of bacteriophages for biocontrol of the kiwifruit canker phytopathogen *Pseudomonas syringae* pv. *actinidiae*. *Appl. Environ. Microbiol.*, 80, 2216–2228.

Fujiwara, A., Fujisawa, M., Hamasaki, R., Kawasaki, T., Fujie, M., Yamada, T. (2011). Biocontrol of *Ralstonia solanacearum* by treatment with lytic bacteriophages. *Appl. Environ. Microbiol.*, 77, 4155–4162.

Gašić, K., Kuzmanović, N., Ivanović, M., Prokić, A., Šević, M., Obradović, A., (2018). Complete genome of the xanthomonas euvesicatoria specific bacteriophage KΦ1, its survival and potential in control of pepper bacterial spot. *Front. Microbiol.*, 9, 2021.

Gill, J. and Abedon, S.T. (2003). Bacteriophage ecology and plants. Feature story, APSnet Feature Artic.

Gill, J. and Hyman, P. (2010). Phage choice, isolation, and preparation for phage therapy. *Curr. Pharm. Biotechnol.*, 11, 2–14.

Hankin, E.H. (1896). L'action bactéricide des eaux de la Jumna et du Gange sur le vibrion du choléra. *Ann. Inst. Pasteur*, 10, 511–523.

Hayward, A.C. (1991). Biology and epidemiology of bacterial wilt caused by *Pseudomonas Solanacearum. Annu. Rev. Phytopathol.*, 29, 65–87.

Hendrick, C.A. and Sequeira, L. (1984). Lipopolysaccharide-defective mutants of the wilt pathogen *Pseudomonas solanacearum. Appl. Environ. Microbiol.*, 48, 94–101.

Henry, M., Lavigne, R., Debarbieux, L. (2013). Predicting in vivo efficacy of therapeutic bacteriophages used to treat pulmonary infections. *Antimicrob. Agents Chemother.*, 57, 5961–5968 [Online]. Available at: https://doi.org/10.1128/AAC.01596-13.

d'Hérelle, F. (1917). Sur un microbe invisible antagoniste des bacilles dysentériques. *C. R. Acad. Sci. Paris*, 165, 373–375.

d'Hérelle, F. (1918). Sur le rôle du microbe filtrant bactériophage dans la dysenterie bacillaire. *C. R. Acad. Sci. Paris*, 167, 970–972.

Hernandez, C.A. and Koskella, B. (2019). Phage resistance evolution in vitro is not reflective of in vivo outcome in a plant-bacteria-phage system. *Evolution*, 73, 2461–2475.

Holtappels, D., Kerremans, A., Busschots, Y., Van Vaerenbergh, J., Maes, M., Lavigne, R., Wagemans, J. (2020). Preparing for the KIL: Receptor analysis of *Pseudomonas syringae* pv. *porri* phages and their impact on bacterial virulence. *Int. J. Mol. Sci.*, 21(8), 2930.

Holtappels, D., Fortuna, K., Lavigne, R., Wagemans, J. (2021). The future of phage biocontrol in integrated plant protection for sustainable crop production. *Curr. Opin. Biotechnol.*, 68, 60–71.

Howard-Varona, C., Hargreaves, K.R., Abedon, S.T., Sullivan, M.B. (2017). Lysogeny in nature: Mechanisms, impact and ecology of temperate phages. *ISME J.*, 11, 1511–1520.

Ibrahim, Y.E., Saleh, A.A., Al-Saleh, M.A. (2017). Management of Asiatic citrus canker under field conditions in Saudi Arabia using bacteriophages and Acibenzolar-S-Methyl. *Plant Dis.*, 101, 761–765.

Iriarte, F.B., Balogh, B., Momol, M.T., Smith, L.M., Wilson, M., Jones, J.B. (2007). Factors affecting survival of bacteriophage on tomato leaf surfaces. *Appl. Environ. Microbiol.*, 73, 1704–1711.

Iriarte, F.B., Obradović, A., Wernsing, M.H., Jackson, L.E., Balogh, B., Hong, J.A., Momol, M.T., Jones, J.B., Vallad, G.E. (2012). Soil-based systemic delivery and phyllosphere in vivo propagation of bacteriophages: Two possible strategies for improving bacteriophage persistence for plant disease control. *Bacteriophage*, 2, 215–224.

Jackson, L.R.E. (1989). Bacteriophage prevention and control of harmful plant bacteria. US4828999A.

Jensen, E.C., Schrader, H.S., Rieland, B., Thompson, T.L., Lee, K.W., Nickerson, K.W., Kokjohn, T.A. (1998). Prevalence of broad-host-range lytic bacteriophages of *Sphaerotilus natans*, *Escherichia coli*, and *Pseudomonas aeruginosa*. *Appl. Environ. Microbiol.*, 64, 575–580.

Jones, J.B., Jackson, L.E., Balogh, B., Obradovic, A., Iriarte, F.B., Momol, M.T., (2007). Bacteriophages for plant disease control. *Annu. Rev. Phytopathol.*, 45, 245–262.

Kaiser, D. and Dworkin, M. (1975). Gene transfer to myxobacterium by *Escherichia coli* phage P1. *Science*, 187, 653–654.

Kan, Y., Lyu, Q., Jiang, N., Han, S., Li, J., Burdman, S., Luo, L. (2020). iTRAQ-based proteomic analyses of the plant-pathogenic bacterium *Acidovorax citrulli* during entrance into and resuscitation from the viable but nonculturable state. *J. Proteomics*, 211, 103547.

Kannan, V., Bastas, K.K., Rajendran, S. (2015). Scientific and economic impact of plant pathogenic bacteria. In *Sustainable Approaches to Controlling Plant Pathogenic Bacteria*, Kannan, V. and Bastas, K.K. (eds). CRC Press, Boca Raton.

Kavagutti, V.S., Andrei, A.-Ş., Mehrshad, M., Salcher, M.M., Ghai, R. (2019). Phage-centric ecological interactions in aquatic ecosystems revealed through ultra-deep metagenomics. *Microbiome*, 7, 135 [Online]. Available at: https://doi.org/10.1186/s40168-019-0752-0.

Kim, M.-H., Park, S.-W., Kim, Y.-K. (2011). Bacteriophages of *Pseudomonas tolaasii* for the biological control of brown blotch disease. *J. Korean Soc. Appl. Biol. Chem.*, 54, 99–104.

Kimmelshue, C., Goggi, A.S., Cademartiri, R. (2019). The use of biological seed coatings based on bacteriophages and polymers against *Clavibacter michiganensis* subsp. *nebraskensis* in maize seeds. *Sci. Rep.*, 9, 17950.

Koczan, J., McGrath, M., Zhao, Y., Sundin, G. (2009). Contribution of *Erwinia amylovora* Exopolysaccharides Amylovoran and Levan to biofilm formation: Implications in pathogenicity. *Phytopathol.*, 99, 1237–44.

Koskella, B. and Taylor, T.B. (2018). Multifaceted impacts of bacteriophages in the plant microbiome. *Annu. Rev. Phytopathology*, 56, 361–380.

Koskella, B., Thompson, J.N., Preston, G.M., Buckling, A. (2011). Local biotic environment shapes the spatial scale of bacteriophage adaptation to bacteria. *Am. Nat.*, 177, 440–451.

Krüger, A. and Lucchesi, P.M.A. (2015). Shiga toxins and stx phages: Highly diverse entities. *Microbiol. Read. Engl.*, 161, 451–462.

Lang, J.M., Gent, D.H., Schwartz, H.F. (2007). Management of *Xanthomonas* leaf blight of onion with bacteriophages and a plant activator. *Plant Dis.*, 91, 871–878.

Lee, L.H., Lui, D., Platner, P.J., Hsu, S.-F., Chu, T.-C., Gaynor, J.J., Vega, Q.C., Lustigman, B.K. (2006). Induction of temperate cyanophage AS-1 by heavy metal–copper. *BMC Microbiol.*, 6, 17.

Leyns, F., De Cleene, M., Swings, J.-G., De Ley, J. (1984). The host range of the genus *Xanthomonas*. *Bot. Rev.*, 50, 308–356.

Lim, J.-A., Lee, D.H., Roh, E., Jung, K., Oh, C., Heu, S. (2013). Biocontrol of *Pectobacterium carotovorum* subsp. *carotovorum* using bacteriophage PP1. *J. Microbiol. Biotechnol.*, 23, 1147–1153.

Lin, D.M., Koskella, B., Lin, H.C. (2017). Phage therapy: An alternative to antibiotics in the age of multi-drug resistance. *World J. Gastrointest. Pharmacol. Ther.*, 8, 162–173.

Mallmann, W. and Hemstreet, C. (1924). Isolation of an inhibitory substance from plants. *Agric. Res.*, 28, 173–179.

Mansfield, J., Genin, S., Magori, S., Citovsky, V., Sriariyanum, M., Ronald, P., Dow, M., Verdier, V., Beer, S.V., Machado, M.A. et al. (2012). Top 10 plant pathogenic bacteria in molecular plant pathology: Top 10 plant pathogenic bacteria. *Mol. Plant Pathol.*, 13, 614–629.

Manyi-Loh, C., Mamphweli, S., Meyer, E., Okoh, A. (2018). Antibiotic use in agriculture and its consequential resistance in environmental sources: Potential public health implications. *Mol. J. Synth. Chem. Nat. Prod. Chem.*, 23(4), 795.

Marchand, P.A. (2017). Basic and low-risk substances under European Union pesticide regulations: A new choice for biorational portfolios of small and medium-sized enterprises. *J. Plant Prot. Res.*, 57, 433–440.

Maxwell, K.L. (2017). The anti-CRISPR story: A battle for survival. *Mol. Cell*, 68, 8–14.

McNeil, D.L., Romero, S., Kandula, J., Stark, C., Stewart, A., Larsen, S. (2001). Bacteriophages a potential biocontrol agent against walnut blight (*Xanthomonas campestris* pv *juglandis*). *N. Z. Plant Prot.*, 54, 220–224.

Meaden, S., Paszkiewicz, K., Koskella, B. (2015). The cost of phage resistance in a plant pathogenic bacterium is context-dependent. *Evol. Int. J. Org. Evol.*, 69, 1321–1328.

Melo, L.D.R., Oliveira, H., Pires, D.P., Dabrowska, K., Azeredo, J. (2020). Phage therapy efficacy: A review of the last 10 years of preclinical studies. *Crit. Rev. Microbiol.*, 1–22.

Mirzaei, M.K. and Maurice, C.F. (2017). Ménage à trois in the human gut: Interactions between host, bacteria and phages. *Nat. Rev. Microbiol.*, 15, 397–408.

Monteiro, R., Pires, D.P., Costa, A.R., Azeredo, J. (2019). Phage therapy: Going temperate? *Trends Microbiol.*, 27, 368–378.

Nagai, H., Miyake, N., Kato, S., Maekawa, D., Inoue, Y., Takikawa, Y. (2017). Improved control of black rot of broccoli caused by *Xanthomonas campestris* pv. *campestris* using a bacteriophage and a nonpathogenic *Xanthomonas* sp. strain. *J. Gen. Plant Pathol.*, 83, 373–381.

Obradovic, A., Jones, J.B., Momol, M.T., Balogh, B., Olson, S.M. (2004). Management of tomato bacterial spot in the field by foliar applications of bacteriophages and SAR inducers. *Plant Dis.*, 88, 736–740.

Peters, D.L., Lynch, K.H., Stothard, P., Dennis, J.J. (2015). The isolation and characterization of two *Stenotrophomonas maltophilia* bacteriophages capable of cross-taxonomic order infectivity. *BMC Genomics*, 16(1), 664.

Pietrzak, U. and McPhail, D.C. (2004). Copper accumulation, distribution and fractionation in vineyard soils of Victoria, Australia. *Geoderma*, 122, 151–166.

Pinheiro, L.A.M., Pereira, C., Frazão, C., Balcão, V.M., Almeida, A. (2019). Efficiency of phage φ6 for biocontrol of *Pseudomonas syringae* pv. *syringae:* An *in vitro* preliminary study. *Microorganisms*, 7, 286.

Pirnay, J.-P., Verbeken, G., Ceyssens, P.-J., Huys, I., De Vos, D., Ameloot, C., Fauconnier, A. (2018). The magistral phage. *Viruses*, 10, 64.

Rabiey, M., Roy, S.R., Holtappels, D., Franceschetti, L., Quilty, B.J., Creeth, R., Sundin, G.W., Wagemans, J., Lavigne, R., Jackson, R.W. (2020). Phage biocontrol to combat *Pseudomonas syringae* pathogens causing disease in cherry. *Microb. Biotechnol.*, 13, 1428–1445.

Rahimi-Midani, A. and Choi, T.-J. (2020). Transport of phage in melon plants and inhibition of progression of bacterial fruit blotch. *Viruses*, 12, 477.

Rahimi-Midani, A., Lee, Y.S., Kang, S.-W., Kim, M.-K., Choi, T.-J. (2018). First isolation and molecular characterization of bacteriophages infecting *Acidovorax citrulli*, the causal agent of bacterial fruit blotch. *The Plant Pathology Journal*, 34, 59–64.

Rahimi-Midani, A., Kim, J.-O., Kim, J.H., Lim, J., Ryu, J.-G., Kim, M.-K., Choi, T.-J. (2020). Potential use of newly isolated bacteriophage as a biocontrol against *Acidovorax citrulli. Arch. Microbiol.*, 202, 377–389.

Ramírez, M., Neuman, B.W., Ramírez, C.A. (2020). Bacteriophages as promising agents for the biological control of Moko disease (*Ralstonia solanacearum*) of banana. *Biol. Control*, 149, 104238.

Rapicavoli, J., Ingel, B., Blanco-Ulate, B., Cantu, D., Roper, C. (2018). *Xylella fastidiosa*: An examination of a re-emerging plant pathogen: *Xylella fastidiosa. Mol. Plant Pathol.*, 19, 786–800.

Ravensdale, M., Blom, T.J., Gracia-Garza, J.A., Svircev, A.M., Smith, R.J. (2007). Bacteriophages and the control of *Erwinia carotovora subsp. carotovora. Can. J. Plant Pathol.*, 29, 121–130.

Rodrigues, C.M., Takita, M.A., Coletta-Filho, H.D., Olivato, J.C., Caserta, R., Machado, M.A., de Souza, A.A. (2008). Copper resistance of biofilm cells of the plant pathogen *Xylella fastidiosa. Appl. Microbiol. Biotechnol.*, 77, 1145–1157.

Rombouts, S., Volckaert, A., Venneman, S., Declercq, B., Vandenheuvel, D., Allonsius, C.N., Van Malderghem, C., Jang, H.B., Briers, Y., Noben, J.P. et al. (2016). Characterization of novel bacteriophages for biocontrol of bacterial blight in leek caused by *Pseudomonas syringae* pv. *porri. Front. Microbiol.*, 7, 279.

Saccardi, A., Gambin, E., Zaccardelli, M., Barone, G., Mazzucchi, U. (1993). *Xanthomonas campestris* pv. *pruni* control trials with phage treatments on peaches in the orchard. *Phytopathol. Mediterr.*, 32, 206–210.

Schroven, K., Aertsen, A., Lavigne, R. (2021). Bacteriophages as drivers of bacterial virulence and their potential for biotechnological exploitation. *FEMS Microbiol. Rev.*, 45, fuaa041.

Schwarczinger, I., Kolozsváriné Nagy, J., Künstler, A., Szabó, L., Geider, K., Király, L., Pogány, M. (2017). Characterization of Myoviridae and Podoviridae family bacteriophages of *Erwinia amylovora* from Hungary – Potential of application in biological control of fire blight. *Eur. J. Plant Pathol.*, 149, 639–652.

Schwartz, D.A. and Lindell, D. (2017). Genetic hurdles limit the arms race between Prochlorococcus and the T7-like podoviruses infecting them. *ISME J.*, 11, 1836–1851.

Seiler, C. and Berendonk, T.U. (2012). Heavy metal driven co-selection of antibiotic resistance in soil and water bodies impacted by agriculture and aquaculture. *Front. Microbiol.*, 3, 399.

Šević, M., Gašić, K., Ignjatov, M., Mijatović, M., Prokić, A., Obradović, A. (2019). Integration of biological and conventional treatments in control of pepper bacterial spot. *Crop Prot.*, 119, 46–51.

Stefani, E., Obradović, A., Gašić, K., Altin, I., Nagy, I.K., Kovács, T. (2021). Bacteriophage-mediated control of phytopathogenic Xanthomonads: A promising green solution for the future. *Microorganisms*, 9, 1056.

Svircev, A., Roach, D., Castle, A. (2018). Framing the future with bacteriophages in agriculture. *Viruses*, 10, 218.

Takahashi, R., Bowatte, S., Taki, K., Ohashi, Y., Asakawa, S., Kimura, M. (2011). High frequency of phage-infected bacterial cells in a rice field soil in Japan. *Soil Sci. Plant Nutr.*, 57, 35–39.

Taylor, P. and Reeder, R. (2020). Antibiotic use on crops in low and middle-income countries based on recommendations made by agricultural advisors. *CABI Agric. Biosci.*, 1, 1.

Tewfike, T.A. and Shimaa, M.D. (2015). Biocontrol of *Xanthomonas axonopodis* causing bacterial spot by application of formulated phage. *Ann. Agric. Sci. Moshtohor.*, 53, 615–624.

Thomas, R.C. (1935). A bacteriophage in relation to Stewarts disease of corn. *Phytopathology*, 25(3), 371–372.

Thompson, D.W., Casjens, S.R., Sharma, R., Grose, J.H. (2019). Genomic comparison of 60 completely sequenced bacteriophages that infect *Erwinia* and/or *Pantoea* bacteria. *Virology*, 535, 59–73.

Tock, M.R. and Dryden, D.T.F. (2005). The biology of restriction and anti-restriction. *Curr. Opin. Microbiol.*, 8, 466–472.

Trotereau, A., Boyer, C., Bornard, I., Pécheur, M.J.B., Schouler, C., Torres-Barceló, C. (2021). High genomic diversity of novel phages infecting the plant pathogen *Ralstonia solanacearum*, isolated in Mauritius and Reunion islands. *Sci. Rep.*, 11, 5382.

Twort, F.W. (1915). An investigation on the nature of ultra-microscopic viruses. *The Lancet*, 186, 1241–1243.

Um, H.Y., Kong, H.G., Lee, H.J., Choi, H.K., Park, E.J., Kim, S.T., Murugiyan, S., Chung, E., Kang, K.Y., Lee, S.-W. (2013). Altered gene expression and intracellular changes of the viable but nonculturable state in *Ralstonia solanacearum* by copper treatment. *Plant Pathol. J.*, 29, 374–385.

Vallad, G.E. and Goodman, R.M. (2004). Systemic acquired resistance and induced systemic resistance in conventional agriculture. *Crop Sci.*, 44, 1920–1934.

Van Cauwenberghe, J., Santamaría, R.I., Bustos, P., Juárez, S., Ducci, M.A., Figueroa Fleming, T., Etcheverry, A.V., González, V. (2021). Spatial patterns in phage-*Rhizobium* coevolutionary interactions across regions of common bean domestication. *ISME J.*, 15, 2092–2106.

Vandamme, E.J. and Mortelmans, K. (2019). A century of bacteriophage research and applications: Impacts on biotechnology, health, ecology and the economy! A century of bacteriophage research and applications. *J. Chem. Technol. Biotechnol.*, 94, 323–342.

Vanneste, J.L. (ed.) (2000). *Fire Blight: The Disease and Its Causative Agent,* Erwinia amylovora. CABI Pub, Wallingford, Oxfordshire.

Vikram, A., Woolston, J., Sulakvelidze, A. (2020). Phage biocontrol applications in food production and processing. *Curr. Issues Mol. Biol.*, 40, 267–302.

Voronina, M.V., Bugaeva, E.N., Vasiliev, D.M., Kabanova, A.P., Barannik, A.P., Shneider, M.M., Kulikov, E.E., Korzhenkov, A.A., Toschakov, S.V., Ignatov, A.N. et al. (2019). Characterization of *Pectobacterium carotovorum* subsp. *carotovorum* bacteriophage PP16 prospective for biocontrol of potato soft rot. *Microbiol.*, 88, 451–460.

Vu, N.T. and Oh, C.-S. (2020). Bacteriophage usage for bacterial disease management and diagnosis in plants. *Plant Pathol. J.*, 36, 204–217.

Wang, X., Wei, Z., Yang, K., Wang, J., Jousset, A., Xu, Y., Shen, Q., Friman, V.-P. (2019). Phage combination therapies for bacterial wilt disease in tomato. *Nat. Biotechnol.*, 37, 1513–1520.

Wei, C., Liu, J., Maina, A.N., Mwaura, F.B., Yu, J., Yan, C., Zhang, R., Wei, H., (2017). Developing a bacteriophage cocktail for biocontrol of potato bacterial wilt. *Virol. Sin.*, 32, 476–484.

Wendling, C.C., Piecyk, A., Refardt, D., Chibani, C., Hertel, R., Liesegang, H., Bunk, B., Overmann, J., Roth, O. (2017). Tripartite species interaction: Eukaryotic hosts suffer more from phage susceptible than from phage resistant bacteria. *BMC Evol. Biol.*, 17, 98.

Woods, T.L., Israel, H.W., Sherf, A.F., Cornell University (1981). Isolation and partial characterization of a bacteriophage of *Erwinia stewartii* from the corn flea beetle, *Chaetocnema pulicaria. Prot. Ecol.*, 3, 229–236

Xin, X.-F., Kvitko, B., He, S.Y. (2018). *Pseudomonas syringae*: What it takes to be a pathogen. *Nat. Rev. Microbiol.*, 16, 316–328.

Yamada, T., Kawasaki, T., Nagata, S., Fujiwara, A., Usami, S., Fujie, M. (2007). New bacteriophages that infect the phytopathogen *Ralstonia solanacearum. Microbiology*, 153, 2630–2639.

Yan, Q., Hu, X., Wang, N. (2012). The novel virulence-related gene nlxA in the lipopolysaccharide cluster of *Xanthomonas citri* ssp. *citri* is involved in the production of lipopolysaccharide and extracellular polysaccharide, motility, biofilm formation and stress resistance. *Mol. Plant Pathol.*, 13, 923–934.

Ye, M., Sun, M., Huang, D., Zhang, Z., Zhang, H., Zhang, S., Hu, F., Jiang, X., Jiao, W. (2019). A review of bacteriophage therapy for pathogenic bacteria inactivation in the soil environment. *Environ. Int.*, 129, 488–496.

Yoshikawa, G., Askora, A., Blanc-Mathieu, R., Kawasaki, T., Li, Y., Nakano, M., Ogata, H., Yamada, T. (2018). *Xanthomonas citri* jumbo phage XacN1 exhibits a wide host range and high complement of tRNA genes. *Sci. Rep.*, 8, 4486.

Zaczek-Moczydłowska, M.A., Young, G.K., Trudgett, J., Fleming, C.C., Campbell, K., O'Hanlon, R. (2020a). Genomic characterization, formulation and efficacy in planta of a Siphoviridae and Podoviridae protection cocktail against the bacterial plant pathogens *Pectobacterium* spp. *Viruses*, 12(2), 150.

Zaczek-Moczydłowska, M.A., Young, G.K., Trudgett, J., Plahe, C., Fleming, C.C., Campbell, K., Hanlon, R.O. (2020b). Phage cocktail containing podoviridae and myoviridae bacteriophages inhibits the growth of *Pectobacterium* spp. under in vitro and in vivo conditions. *PLoS One*, 15, e0230842.

Zhao, Y., Ye, M., Zhang, X., Sun, M., Zhang, Z., Chao, H., Huang, D., Wan, J., Zhang, S., Jiang, X. et al. (2019). Comparing polyvalent bacteriophage and bacteriophage cocktails for controlling antibiotic-resistant bacteria in soil-plant system. *Sci. Total Environ.*, 657, 918–925.

Zhou-qi, C., Bo, Z., Guan-lin, X., Bin, L., Shi-wen, H. (2016). Research status and prospect of *Burkholderia glumae*, the pathogen causing bacterial panicle blight. *Rice Sci.*, 23, 111–118.

9

Microbiome-assisted Agriculture: Current Knowledge and Future Directions

Ke YU[1], Hongwei LIU[2], Wei ZHONG[3] and Ioannis A. STRINGLIS[4]
[1] *Henan University, China*
[2] *Hawkesbury Institute for the Environment, Western Sydney University, Australia*
[3] *Nanjing Agricultural University, China*
[4] *Department of Biology, Utrecht University, Netherlands*

9.1. Introduction

The global population has been steadily expanding over recent decades and is expected to reach 9.7 billion in the next 20 years (Roser 2013). This prospect is expected to pose great challenges to agriculture, to meet the ever-growing food demands. Most scientific efforts in the last century aimed to maintain a good level of agricultural productivity by generating high-yield plant varieties or by introducing chemical pesticides and fertilizers in agricultural practice. However, these approaches are not sustainable in the long run. One reason is that many of the high-yield varieties require a lot of water or nutrient inputs, or they are susceptible to diseases caused by pathogens or insect infestations (Boyd et al. 2013). Another reason is that the extensive use of chemicals to increase productivity has a negative impact on the environment and human health and also exerts pressure on pathogens that develop resistance to the most common drastic compounds used to

Biocontrol of Plant Disease,
coordinated by Claire PRIGENT-COMBARET and Bernard DUMAS. © ISTE Ltd 2022.

suppress them (Mann et al. 2009; Jiao et al. 2012; Nicolopoulou-Stamati et al. 2016). It is therefore urgent to improve crop productivity using alternative solutions. It is also important that these solutions are efficient, sustainable and have a minimum impact on the environment and human health (Gomiero et al. 2011). One such approach is to exploit the plant beneficial traits of microbes that are associated with plants, collectively known as plant microbiome, with the aim of providing agriculture with an eco-friendly approach to boost plant health and yield (Berendsen et al. 2012; Finkel et al. 2017; Berg et al. 2020). To achieve this, it is critical to know what determines the composition and functions of plant-associated microbial communities, how plants communicate with their microbiomes, what the microbial services provided to the plants are and what the means to engineer microbial consortia that can alleviate stress and improve crop yields in field conditions are.

9.1.1. *The plant microbiome*

The soil, where plant roots grow, is a complex ecosystem with millions of microbial inhabitants residing in really high densities per gram of soil (Berendsen et al. 2012). These microbes, together with the functions they encode, are known as the microbiome (Figure 9.1). The soil microbiome consists of fungi, bacteria, protists, algae, archaea and viruses (Berg et al. 2020). Plants can modulate the soil microbiome and attract a subset of its members which will then colonize different plant organs and tissues. It was at the beginning of the last century, when Lorentz Hiltner discovered that there is a complex interplay between plant roots and the microbes around them. He suggested that plants can affect microbial density and attract beneficial bacteria in the thin soil layer surrounding the roots, termed the rhizosphere, and called this phenomenon the "rhizosphere effect" (Hartmann et al. 2008; Bakker et al. 2020). This communication between plant roots and the root-associated microbiome is mediated by the release of different molecules from the plants and/or the production of diverse microbial metabolites (Sasse et al. 2018; Stringlis et al. 2018b; Pascale et al. 2020). The plants release up to 40% of photosynthetically fixed carbon in the rhizosphere in the form of organic acids, amino acids, sugars and secondary metabolites, turning the rhizosphere into a hotspot of microbial activity and favoring microbe–microbe and plant–microbe interactions.

The rhizosphere microbiome is a subset of the soil microbiome, and is mostly a less diverse but more dense microbial community (Bulgarelli et al.

2013; Bakker et al. 2020). The microbial communities proliferating in the inner plant tissues (endosphere) or different plant organs such as leaves and flowers (phyllosphere) are even more selected subsets of the soil/seed/rhizosphere microbiomes, suggesting that plant processes are involved in the selection of the microbes establishing in niches throughout the plant body (Bulgarelli et al. 2013; Liu et al. 2017a; Trivedi et al. 2020; Stassen et al. 2021).

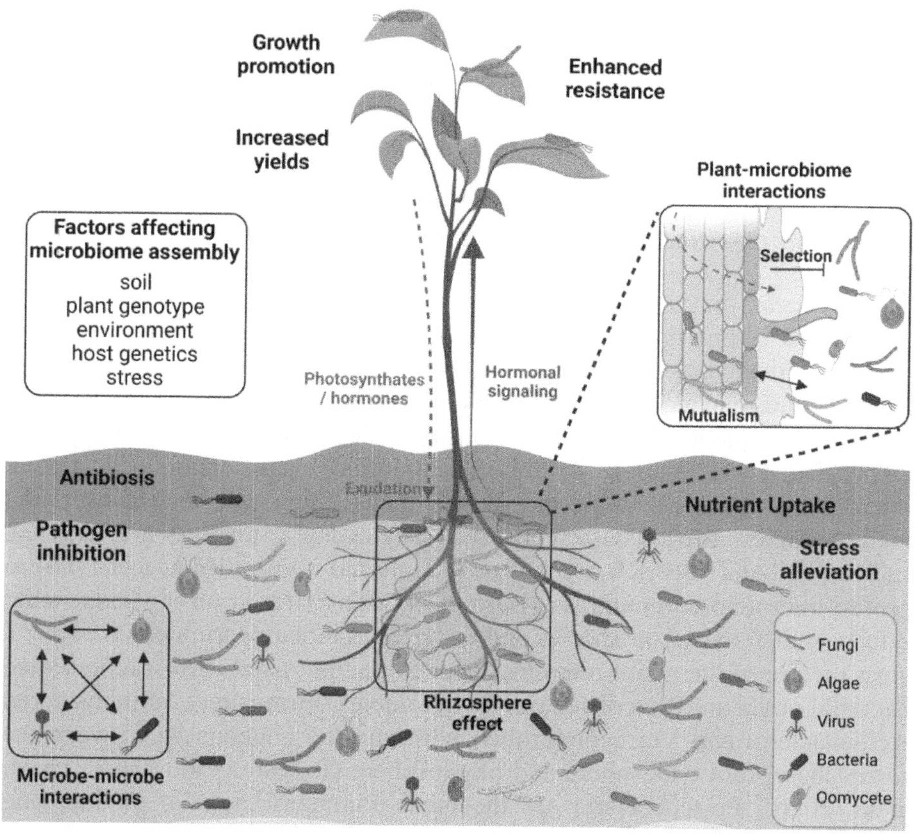

Figure 9.1. *The plant microbiome. This figure was created using Biorender.com. For a color version of this figure, see www.iste.co.uk/prigentcombaret/biocontrol.zip*

COMMENT ON FIGURE 9.1.– *Numerous and diverse microbes live around plant roots including bacteria, fungi, algae, archaea, viruses and oomycetes. The composition of these communities is affected either by plant activity or microbe–microbe interactions. Plants select their associated microbiomes via the exudation of various metabolites or the activity of hormones in the rhizosphere (rhizosphere effect). In this case, some microbes can establish mutualistic interactions with plants, while others are inhibited by plant exudation and defense responses (plant–microbiome interactions). The selected microbes can provide their hosts with various beneficial services such as nutrient uptake, antibiosis and pathogen inhibition, stress alleviation and induction of enhanced resistance against pathogens, leading to increased yield and growth. This can happen via the modulation of hormonal signaling of the plants by the plant-associated microbiome. Apart from plant activity, other important factors with a role in microbiome assembly and activity are the soil, the plant genotype, environmental perturbations and conditions of stress (e.g. pathogen or insect attack).*

9.1.2. *Composition of the plant microbiome*

Advances in next-generation sequencing, the development of bioinformatic pipelines and the creation of more and more complete microbial databases combined with the improvement of approaches to generate and manage microbial isolate collections have allowed the detailed description of plant microbiome composition (Sczyrba et al. 2017; Zhang et al. 2021). Most information about the taxonomic classification of different microbiome members is provided by studies using amplicon sequencing methodologies. In this case, conserved phylogenetic markers are targeted, such as 16S and 18S rRNA genes or the internal transcribed spacer (ITS) to profile microbial communities (Berg et al. 2020). Based on the sequencing efforts in different plant species and different plant compartments, we have a clear picture of the major microbial taxa colonizing plants. Most represented bacterial phyla are Proteobacteria, Bacteroidetes, Firmicutes, Actinobacteria, Acidobacteria and Verrucomicrobia, with varying abundances between the rhizosphere, root endosphere and phyllosphere (Pascale et al. 2020; Trivedi et al. 2020). Proteobacteria are the most dominant bacterial phylum and represent around 50% of the total bacterial community irrespective of plant species or plant tissue (Bai et al. 2015; Fitzpatrick et al. 2018; Carlstrom et al. 2019). Most fungi colonizing plants belong to phyla Ascomycota and Basidiomycota (de Souza et al. 2016; Almario et al. 2017; Bergelson et al. 2019). In the case of archaea, most represented archaeal taxa associated with plants belong to the phyla Thaumarchaeota, Crenarchaeota and Euryarchaeota (Taffner et al. 2018; Jung et al. 2020). Protists represent an important fraction of the plant microbiome with Oomycota and Cercozoa representing important members of those that have variable effects on plant health (Sapp et al. 2018; Gao et al. 2019; Xiong et al. 2020). Viruses (i.e. pathogen-specific phages) are also important players in the plant microbiome and a recent study demonstrated the role of phages in controlling a bacterial wilt disease in tomato (Wang et al. 2019b). Finally, different algal genera such as *Anabaena* and *Chlorella* spp. are also isolated from many plant niches and seem to have the potential to promote plant health and growth (Lee and Ryu 2021). All the above-mentioned microbiome members, despite their varying sizes and colonization potential, and their morphological, biological and phylogenetic diversity, have their own means to support plant growth and health.

9.1.3. *Contribution of the microbiome to plant fitness*

Plant microbiome members are as diverse as they are numerous. They can be detrimental, neutral or beneficial for their plant hosts. Devastating plant pathogens are members of bacterial (e.g. *Ralstonia solanacearum*, *Pseudomonas syringae*), fungal (*Fusarium* sp., *Magnaporthe* sp., etc.) and oomycete (*Pythium* sp., *Phytophthora* sp., etc.) microbiomes associated with plants (Dean et al. 2012; Mansfield et al. 2012; Kamoun et al. 2015). Nevertheless, plants mostly maintain a healthy state in nature due to the activity of the microbiome members that provide their hosts with important services and counteract the negative impact of pathogens (Figure 9.1). The so-called beneficial microbiome assists plants in nutrient and water uptake, is involved in stress amelioration, can directly inhibit the growth of pathogens or indirectly boost the plant immune system of their hosts. The results of all these services are plant growth promotion and yield improvement (Pieterse et al. 2014; Rolli et al. 2015; Finkel et al. 2017; Hassani et al. 2018; Bakker et al. 2020; Pascale et al. 2020). In the following sections, we will outline how the microbiome can contribute to the optimal growth of plants and the maintenance of homeostasis under conditions of stress.

9.2. Microbiome-mediated benefits for plants

The beneficial associations formed between plants and microbes are widespread in nature. A well-known example is the unique symbiosis formed between leguminous plants and rhizobia; plants provide living space and carbon sources to accommodate rhizobia that can provide readily usable nitrogen through biological fixation (Oldroyd et al. 2011). Another important symbiotic relationship is formed between most terrestrial plants and mycorrhiza fungi, where plants provide carbon in the form of lipids to fungi and fungi, in turn, improve the phosphorus uptake in host plants (Genre et al. 2020). The establishment of such symbiotic associations typically involves the development of specialized physiological structures in the host–microbe interface, such as rhizobia-triggered nodules in legumes and arbuscules formed by mycorrhiza fungi inside plant roots (Gage 2004; Genre et al. 2008). Selected microbes, that do not engage with host plants in such a way, can still confer a wide range of beneficial effects on host plants via various strategies. These microbes, namely plant growth-promoting bacteria and fungi, can aid in nutrient and water uptake, inhibit pathogens directly via resource competition or indirectly via elicitation of immunity, and enhance tolerance against abiotic stresses (Lugtenberg and Kamilova 2009; Pieterse et al. 2014). The molecular mechanisms underlying the beneficial plant–microbe associations have been extensively investigated and numerous host genetic factors and microbial molecules are identified as determining factors to build such relationships. Till today, most of our knowledge on this topic comes from the studies that only considered binary associations between one plant and one microbial species. In nature, however, plants simultaneously interact with highly diverse microbial communities, and emerging evidences suggest that microbiomes as a whole can largely determine plant health (Berendsen et al. 2012).

9.2.1. *Disease suppressive soils*

One of the most well-known examples of microbiome-mediated benefits to the host plants is the creation of disease suppressive soils. Plants growing on this kind of soil exhibit enhanced disease resistance against specific soilborne pathogens (Weller et al. 2002). In a monoculture wheat field, the severity of take-all disease caused by the pathogenic fungus *Gaeumannomyces graminis* var. *tritici* gradually increases in the first few years, and then declines substantially (Weller et al. 2002). This phenomenon only occurs in monoculture fields but was not observed when crop rotation is applied, and has been demonstrated to be associated with specific enrichment of *Pseudomonas* spp. in wheat rhizospheres that are able to inhibit the soilborne pathogen via production of antifungal 2,4-diacetylphloroglucinol (Weller et al. 2002). One of the first studies aiming to unearth the mechanism behind this unique soil property appeared in 2011. The researchers applied metagenomics combining culture-dependent functional analysis to reveal the differences in the rhizosphere microbiome composition of sugar beet plants that were grown on disease suppressive soil and disease inducive soil (Mendes et al. 2011). They observed an enrichment of γ-Proteobacteria equipped with non-ribosomal peptide synthetases in the rhizosphere of sugar beet seedlings grown in disease suppressive soil, which is involved in the resistance against the root fungal pathogen *Rhizoctonia solani* (Mendes et al. 2011). In a follow-up study by the same group, a microbial consortium consisting of *Chitinophaga* and *Flavobacterium* was identified in the endophytic microbiome of sugar beet that can protect the host against the same pathogen (Carrion et al. 2019). Furthermore, they identified additional microbial genetic traits that are essential for the protective effect exerted by the microbiome (Carrion et al. 2019).

9.2.2. *Induced resistance and soilborne legacy*

The mechanisms underlying microbiome-mediated disease resistance can be diverse. Apart from direct antagonistic effect on pathogens, selected members can also prime plant immunity for boosted immune responses against potential attacks (Pieterse et al. 2014). In this case, the inhibitory effect of the microbiome on pathogens is not restricted to the sites where the microbiome assembles, but is extended to systemic host tissues where potential future attacks will happen (Pieterse et al. 2014). This form of immunity is known as induced systemic resistance, is effective against multiple pathogens and insects and is observed in different plant species (Pieterse et al. 2014). Several studies have reported that systemic signals are generated and transmitted between root and shoot, which serve as triggers of microbiome reconstruction and functional optimization of the microbiome (Stassen et al. 2021). Foliar pathogen infections can significantly alter the root microbiome composition (Berendsen et al. 2018; Yuan et al. 2018). Interestingly, these disease-induced compositional shifts in root microbiome ultimately led to a systemically enhanced disease resistance in aboveground plant tissues (Berendsen et al. 2018; Yuan et al. 2018), thereby building up a perfect regulatory loop consisting of shoot, root and microbiome. Importantly, this resistance can be passaged to the upcoming generation of plants that will grow in the same soil (Berendsen et al. 2018; Yuan et al. 2018). In this way, plants can modify the composition of their microbiome to create a soilborne legacy, which confers inheritable advantages to their offspring (Bakker et al. 2018). However, despite numerous cases of immunity-eliciting microbes, there are few reports on whether and how the whole microbiome modulates the development of a mature immune system in plants. In-depth investigation of the immune robustness of plants in the presence and absence of a representative microbiome is required to answer these questions.

9.2.3. *Nutrient stress and microbiome*

Nutritional deficiencies in essential elements, or inappropriate climate conditions such as drought and heat, or other environmental stresses such as salinity and heavy metal pollutants can severely affect plant fitness (Mittler 2006). The microbiome can help the plants deal with those abiotic stresses as well (Rolli et al. 2015). A recent study nicely illustrated the mechanism of microbiome-mediated root iron uptake in the model plant species *Arabidopsis thaliana*. Different microbiome members share a functionally redundant iron-mobilization activity that is elicited by plant-derived molecules and required for improved iron nutrition in host plants (Harbort et al. 2020). A similar beneficial effect of the microbiome against abiotic stresses was observed in another study, not only in defined laboratory conditions but also in the field. In this case, researchers dissected the root microbiome composition in a diverse panel of field-grown rice varieties and identified that the *indica* varieties with superior nitrogen-use efficiency were enriched in more microbial taxa with nitrogen metabolism functions than *japonica* varieties, which are less efficient in nitrogen-use efficiency (Zhang et al. 2019). Moreover, reinoculation of *indica*-enriched microbes, but not *japonica*-enriched microbes, significantly improved the growth of an *indica* variety, compared to microbe-free rice (Zhang et al. 2019), indicating that microbiome-mediated enhancement of resistance against stresses, at least in some cases, may require not only the presence but also the adapted composition of the microbiome.

9.2.4. *Plant immune system and microbiome*

Plants often encounter combinational stresses simultaneously (Mittler 2006). Therefore, they need to prioritize different stress responses for fitness optimization. In recent years, a regulatory role of microbiomes in such processes was revealed. Plant defense is generally considered as an energy-demanding process that can affect growth, a phenomenon that is referred to as growth-defense tradeoff (Huot et al. 2014). Microbiome members all possess similar immunogenic elicitors that can be recognized by the plant immune system and trigger immune responses (Teixeira et al. 2019; Yu et al.

2019; Stringlis and Pieterse 2021). To achieve successful colonization, root microbiome members have evolved various strategies to overcome plant defense. Two recent studies using defense-mediated root growth inhibition as an index for immunity have shown that different root microbiome members have varying capabilities of host immune suppression (Ma et al. 2021; Teixeira et al. 2021). Colonization by an immune-suppressive microbiome would render the plant roots more vulnerable to colonization by non-suppressive members or even pathogens (Ma et al. 2021; Teixeira et al. 2021). In this scenario, the root microbiome that generally consists of suppressive and non-suppressive members, functions as a buffer for plant defense and defense-associated growth reduction. The growth-defense tradeoff modulated by the root microbiome was also observed in aboveground plant tissues. Under suboptimal light, the plants can regulate the root microbiome composition via a thus far unidentified long-distance signal (Hou et al. 2020). Microbiome reconstruction can reciprocally improve the growth parameters of plants grown under low light conditions via the downregulation of microbiome-elicited immune responses in aboveground tissues (Hou et al. 2020), again indicating a microbiome–root–shoot regulatory loop that can help plants optimizing stress responses. The interplay between defense and other types of stress responses were also reported to be modulated by the microbiome. In *Arabidopsis*, a root microbiome was shown to promote the uptake of nutrients, such as iron and phosphorus, via the activation of nutritional deficiency responses (Zamioudis et al. 2015; Castrillo et al. 2017). Moreover, these responses are found to be associated with a clear suppression of plant immunity, which would definitely favor the colonization or proliferation of selected members of the plant microbiome (Castrillo et al. 2017; Harbort et al. 2020). The intriguing regulatory role of the plant microbiome in the interactions between different plant biological processes suggests that there is an intrinsic link between microbiome and plant stress-responsive machineries.

9.3. Chemical cues derived from plants and microbes guide microbiome assembly

Studies in the last decade have demonstrated that microbiome composition can be affected by many different factors such as the soil type, environmental perturbations, host genetic factors and also conditions of stress (Figure 9.1) (Mendes et al. 2011; Bulgarelli et al. 2012, 2015; Lundberg et al. 2012; Chaparro et al. 2013; Carvalhais et al. 2015; Edwards et al. 2015; Lebeis et al. 2015; Chapelle et al. 2016; Dombrowski et al. 2017;

Hacquard et al. 2017; Haney et al. 2017; Zhang et al. 2017; Berendsen et al. 2018; Duran et al. 2018; Fitzpatrick et al. 2018; Hassani et al. 2018; Kwak et al. 2018; Mendes et al. 2018; Stringlis et al. 2018a; Cordovez et al. 2019; Cotton et al. 2019; Huang et al. 2019; Liu et al. 2019b, 2021; Pérez-Jaramillo et al. 2019; Wang et al. 2019b; Zhang et al. 2019; Snelders et al. 2020; Trivedi et al. 2020). The microbiome is vital for plant health and yield, which raises a possibility that a microbiome could be manipulated or engineered for desired biological functions. Knowledge on the principles of microbiome assembly is the premise for ultimately achieving this goal in agriculture. One of the main obstacles in understanding the principles of the microbiome assembly process is the accuracy of compositional profiling. Advances in high-throughput sequencing technology have largely overcome this issue and brought prosperity to the microbiome research field (Gray et al. 2015). As for plant microbiomes, factors that contribute to the formation of a functional microbiome are incredibly diverse, since the determining effects on microbiome composition can be attributed to plants, microbes, the environment or the interactions amongst them. Most environmental factors affecting microbiome assembly, if not all, act not only on the microbes but also on the plants and the interactions between them. For the sake of simplicity, here we mainly focus on the plant and microbe-derived assembly cues that have been proven to be directly functional in restructuring microbiome composition.

9.3.1. *Microbiome assembly mediated by host plants*

Plant-associated microbes – pathogenic, beneficial or commensal – all heavily rely on the nutrients released by host plants such as sugars and lipids. Emerging evidence suggests that the plant immune system, the major detection and responding system that plants have evolved to cope with colonizing microorganisms, plays an active role in maintaining microbial homeostasis (Teixeira et al. 2019; Yu et al. 2019; Stringlis and Pieterse 2021). Extensive work focusing on the dissection of the host genetic basis in the assembly of plant microbiomes has yielded fruitful insights into this. Plant traits including compartments (Bulgarelli et al. 2012; Lundberg et al. 2012; Edwards et al. 2015), genetic variations (Bulgarelli et al. 2012; Peiffer et al. 2013), stress-responsive mechanisms (Castrillo et al. 2017; Harbort et al. 2020) and developmental stages (Chaparro et al. 2014; Beilsmith et al. 2021) have been implicated in the microbiome assembly process, revealing specific microbial adaptation to host traits. Most of these works mainly spotted the correlation; however, a causal link between those factors and the compositional shifts in the microbiome is still missing. So far, most plant-derived microbiome-modulating factors are chemical molecules. It was reported that the plant physical structures, for example root hairs, can determine the composition of root microbiome (Robertson-Albertyn et al. 2017). However, developmental defects may have pleiotropic effects on various biological processes, and are often accompanied by changes in the host metabolome profiles as well as in the magnitudes of defense responses. Therefore, more detailed investigations are required to dissect the exact functions of morphological changes in microbiome assembly.

Plant metabolites are indispensable in plant growth, development and resistance against various biotic/abiotic stresses as demonstrated by using mutants devoid of a certain class of phytochemicals (Piasecka et al. 2015; Wang et al. 2019a). In recent years, a clear causality has been established between plant metabolism and microbiome composition (Pascale et al. 2020). Many metabolites, such as lignin, can be integrated into physical structures such as cell walls to support plant growth and development (Tobimatsu and Schuetz 2019). Many other metabolites can be released into the environment in the form of root exudates such as coumarins (Tsai and Schmidt 2017; Stringlis et al. 2019) or gaseous volatiles such as terpenes (Thimmappa et al. 2014). The metabolome profiles of host plants can change according to their developmental stages, or in response to the stresses (Piasecka et al. 2015; Wang et al. 2019a). Therefore, it is expected that alteration of the metabolome would affect microbiome assembly in both the plant interior or exterior via selectively inhibiting or promoting the resident microbes. Physical layers formed by plant metabolites such as cuticle, suberin and lignin are important components of pre-formed defense barriers in plants (Schreiber 2010; Serrano et al. 2014; Roppolo et al. 2014). For example, upon perception of microbes, plant roots increased the callose deposition in the epidermis to halt the potentially invading pathogens (Millet et al. 2010; Luna et al. 2011). However, few studies indicate that failure in the formation of such barriers in host plants has a direct impact on the microbiome assembly. Previously, it was shown that *Populus alba*, which carries a mutation in cinnamoyl-CoA reductase, a key enzyme in the lignin biosynthesis, possesses a distinct endophytic microbiome (Beckers et al. 2016). The compositional shifts in this case are likely due to the altered ferulic acid level in this mutant, which favors more microbes with strong metabolic capability of ferulic acid (Beckers et al. 2016). Nonetheless, a recent study showed that an *Arabidopsis* carrying mutations in two receptor-like kinases ERK1 and TIC1 exhibited disrupted Casparian strips and enhanced suberin lamellae formation, as well as a significantly altered microbiome (Durr et al.

2021). In another study, *Arabidopsis* mutants with modified root diffusion barriers assembled different bacterial communities compared to wild-type plants and the authors demonstrated the dynamic interplay between different microbiome members and the formation of endodermal suberin (Salas-Gonzalez et al. 2021). Casparian strips and suberin lamellae were proven to function as endodermal barriers preventing the diffusion of microbial immune elicitors from root exterior to stele, therefore compartmentalizing immune responses to ensure robust immune output against invading pathogens while mild immune output towards non-invading microbes was observed (Zhou et al. 2020). Therefore, the different microbiome composition in mutants with abnormal Casparian strips and suberin lamellae could be a result of compromised immune-gating system.

A fundamental role of plant metabolism in stress resilience has been well documented. Plant defense hormones, mainly salicylic acid, jasmonic acid and ethylene, were shown to affect the microbiome composition (Bodenhausen et al. 2014; Lebeis et al. 2015; Carvalhais et al. 2017). However, considering the impact of defense hormones on overall plant fitness, more in-depth studies are needed to reveal the underlying mechanisms. Besides, plants synthesize specialized metabolites to cope with multiples stresses such as pathogens, insects, drought, salinity and nutrient deficiency. Immune-related metabolites, such as camalexin, glucosinolates (GLS), triterpenes and benzoxazinoids (BXs), possess antimicrobial activities, which directly drive the changes of plant microbiome. Tryptophan-derived camalexin is a sulfur-containing indolic compound that is mainly found in Cruciferous plants including *Arabidopsis* (Glawischnig 2007). Interestingly, a study showed that camalexin can be released as part of the root exudate blend into the rhizosphere and causes compositional shifts in root microbiome by affecting microbial sulfatase activities (Koprivova et al. 2019). A similar phenomenon was found for another class of sulfur-containing antimicrobial metabolites, GLS (Bressan et al. 2009; Hiruma et al. 2016). The biosynthetic route of mevalonate-derived triterpene was fully elucidated in *Arabidopsis* recently (Huang et al. 2019). Two types of triterpenes, arabidin and thalianin, can modulate the assembly of root microbiome, as different microbes exhibited varying capabilities in the metabolism and utilization of these compounds (Huang et al. 2019). Terpenes can be released as volatiles as well and a study showed that an alteration of terpene emission could modulate the composition of floral microbiome (Boachon et al. 2019). In cereals, such as wheat and maize, BXs are major chemical weapons against phytopathogenic microbes and insect attack (Zhou et al. 2018). A recent study showed that BXs-derived molecules, such as 6-methoxy-benzoxazolin-2-one, altered the composition of maize root microbiome, subsequently leading to enhanced insect resistance that is dependent on jasmonic acid signaling (Hu et al. 2018). The biosynthesis of BXs affects the production of diverse metabolites as well, especially flavonoids, which were shown to affect the assembly of root microbiome in maize (Cotton et al. 2019; Kudjordjie et al. 2019; Cadot et al. 2021). These results suggest that plant defense metabolites might act as another layer of the immune-gating system that can keep the microbiome in check.

Apart from biotic stresses, abiotic stresses can also significantly affect plant metabolism. Phenylalanine-derived coumarins represent metabolites that are synthesized and secreted by plant roots under iron limiting conditions (Tsai and Schmidt 2017; Stringlis et al. 2019). Feruloyl-CoA 6'-hydroxylase 1 is the committed enzyme that is indispensable for coumarin biosynthesis in *Arabidopsis* (Schmid et al. 2014). Microbiome dissection and reconstitution assays using *f6'h1* demonstrated that coumarins could affect microbiome assembly potentially via the generation of reactive oxygen species (ROS), that directly eliminated the growth of certain microbes (Stringlis et al. 2018a; Voges et al. 2019). Phosphate deficiency was shown to drive microbiome assembly through the master transcriptional regulator of phosphate starvation responses, PHR1 (Castrillo et al. 2017). This modulation is closely linked to the dual regulatory roles of PHR1 in both nutritional stress and defense, as activation of PHR1 by phosphate deficiency can directly repress immunity (Castrillo et al. 2017). Notably, it is implicated that phosphate deficiency could also trigger the biosynthesis and secretion of coumarins (Tsai and Schmidt 2017). In maize, nitrogen deprivation could enhance flavone production, resulting in the enrichment of Oxalobacteraceae in the root microbiome that facilitates plant nitrogen uptake via the formation of more lateral roots (Yu et al. 2021b). In sorghum, drought stress could restructure the root microbiome and lead to an enrichment of monoderm bacteria that typically have thicker cell walls (Xu et al. 2018). This compositional shift is correlated with the alteration of plant-derived metabolites, such as glycerol-3-phosphate, in response to drought (Xu et al. 2018).

Other plant-derived chemicals can exert an effect on the composition of the microbiome as well. In two recent studies, a prominent role of plant-derived ROS in microbiome assembly was revealed. Researchers have identified that *Arabidopsis* mutants defective in NADPH oxidase assembled a distinct rhizosphere microbiome that enriched in *Pseudomonas* compared to wild-type plants (Song et al. 2021), while the leaf compartment showed the enrichments of bacteria belonging to the class Gammaproteobacteria, which likely contains opportunistic pathogens (Pfeilmeier et al. 2021). Together, both studies emphasized a vital function of NADPH oxidases in the prevention of plant microbiome dysbiosis. The mechanisms of ROS-mediated microbiome assembly could be diverse. ROS function as important signaling molecules in a plethora of stress signaling pathways (Mittler 2017). Deficiency in ROS generation severely affects the plant immune output (Ngou et al. 2021; Yuan et al. 2021), thereby allowing the uncontrollable growth of certain microbes. Besides, as an oxidizing agent, ROS can directly act on a broad range of microbes to restrict or eliminate their growth (Fang 2011).

9.3.2. *Microbiome assembly mediated by microbe–microbe interactions*

In nature, a microbiome typically contains extremely diverse microbial species which consistently interact with each other (Hassani et al. 2018). The interactions between co-inhabiting microbes can be cooperative or competitive, occur at the inter- or intra-taxa level, and are generally studied in well-controlled systems that involve a small number of microbial species. Understanding how microbe–microbe interactions can affect microbiome assembly, however, is challenging: the number of interactions that happen simultaneously in a natural system is enormous. Moreover, the observed microbial effect on a microbiome could also come from the host plants after perception of certain microbes. Therefore, apart from traditional microbiome dissections in which host plants are present, many research works have started constructing synthetic microbiomes using microbes isolated from the host microbiome. In this way, the microbiome composition can be easily designed based on experimental purposes and the presence of host plants becomes a variable in the analysis. Microbial species could be implemented (bottom-up) or dropped out (top-down) during the construction process, to investigate the function of certain microbial species in a fixed microbiome (Finkel et al. 2020). An array of bioinformatic approaches (such as network analysis of correlated taxa) were also developed to get a glimpse into the microbial interactions in a given microbiome and get insights on the identifications of microbial hubs or keystone species that are central determinants of the microbiome structure (Kroll et al. 2017; Banerjee et al. 2018).

The cooperative interactions among microbes are essential to confer fitness benefits. Very often, mono bacterial species or a large bacterial community can form biofilm, a protective dense polymeric matrix that is resistant to environmental stresses, on specific plant sites (Danhorn and Fuqua 2007). Within such a matrix, it is convenient for microbes to ensure the maxima of functional efficiency that cannot be achieved by singletons. A study on *Arabidopsis* plants that were infected with foliar pathogens revealed an exceptional enrichment of a bacterial consortium consisting of three species of *Xanthomonas*, *Stenotrophomonas* or *Microbacterium* in the root microbiome (Berendsen et al. 2018). Notably, when grown together, the three bacterial species were able to form biofilm in vitro and exhibited significantly stronger growth-promoting and immune-eliciting capabilities on host plants than as single species (Berendsen et al. 2018). The recruitment of such a plant-beneficial consortium is likely mediated by plant-derived cues as a result of defense activation, as well as microbial interactions indicated by their close relationships (Yu et al. 2021a). Poor nutrient availability in environments is typically a limiting factor for microbial growth, and many microbes have evolved to reciprocally exchange metabolites to cope with this issue (Hassani et al. 2018). A study using a synthetic defined growth medium and microbiomes derived from multiple sources observed that the microbiomes, despite their initial compositional differences, exhibited a convergent structure after successive passaging from different growth media (Goldford et al. 2018). Moreover, using mathematical modeling, researchers revealed a predominant role of non-specific cross-feeding with metabolites in structuring a stable microbiome (Goldford et al. 2018), indicating that microbial cooperation can largely determine the outcome of microbiome assembly.

Plant-associated microbes also compete with each other directly or indirectly for resource or niche occupancy (Lugtenberg and Kamilova 2009). Well-known examples are microbial biocontrol agents against plant pathogens, which can inhibit pathogen growth via various mechanisms including but not limited to nutrient sequestrators (Gu et al. 2020), antimicrobial compounds (Mazurier et al. 2009; Stringlis et al. 2018b), effector proteins (Snelders et al. 2020) and volatile emissions (Raza et al. 2020). In many cases, the competition via the secretion of such molecules is not directional or targeted. Therefore, an overall impact on co-inhabiting microbes is expected. Interestingly, the competition can also occur at a community level. A comprehensive study using a representative core set of microbes isolated from *Arabidopsis* demonstrated that the presence of a bacterial microbiome is necessary to prevent the disease caused by fungi or oomycetes present in the soil (Duran et al. 2018), an observation linked to the previously described soil fungistasis (Bakker et al. 2020). The interkingdom inhibitory effect conferred by the bacterial microbiome is mediated by the redundant antagonistic activities against fungi/oomycetes shared by a large group of bacterial species (Duran et al. 2018). Interestingly, it is also shown that an adapted plant fungal pathogen, *Verticillium dahliae*, can use two effector proteins, VdAve1 and VdAMP2, to directly suppress antagonistic bacteria in the root microbiome (Snelders et al. 2020). Similarly, a bioinformatic analysis of bacterial adaptation features to plants identified Hyde1, an effector protein secreted by the type 6 secretion system of *Acidovorax* strains, capable of inhibiting various plant-associated bacteria (Levy et al. 2018). Together, these results extend the functions of microbial effector proteins beyond their extensively studied in planta virulence.

9.4. Plant and soil microbiome engineering

Understanding how the plant manipulates its microbiome spatially and temporally under biotic and abiotic stresses will now make it possible to identify microbial traits that could enhance the capacity of plants in stress resilience and develop promising strategies for plant/soil microbiome engineering to achieve microbiome-assisted agriculture. The evidence that crop varieties with better disease resistance and/or nutrient utilization efficiency harbor superior microbial traits in their rhizosphere suggest that plant microbiome can prominently contribute to a particular phenotype of plant adaptation to a stressful environment (Kwak et al. 2018; Mendes et al. 2018; Zhang et al. 2019). All these examples confirm that there are intrinsic microbial traits in the plant microbiome that can be manipulated to microbially breed high-yielding and nutrient-efficient crop varieties (Wei and Jousset 2017).

9.4.1. *"Cry for help" strategy for the plant to accumulate beneficial microbes*

Plants have been found to attract beneficial microbes from the surrounding soils to counter insect- and pathogen attacks, a phenomenon known as "cry for help" (Berendsen et al. 2018; Carrion et al. 2019; Liu et al. 2020, 2021). Those microbes responding to the call of stressed plants can prime plant signaling pathways and/or inhibit growth or virulence of soilborne pathogens (e.g. *Fusarium* spp. and *R. solani*) (Berendsen et al. 2018; Carrion et al. 2019; Liu et al. 2021). As such, "cry for help" can be a critical strategy for the survival of plant individuals and their offspring when they encounter biotic stresses (Bakker et al. 2018). Recent research combining metagenomics with culture-dependent methods has revealed that disease-attracted beneficial microbe(s) in the rhizosphere/root endosphere/ phyllosphere included *Chitinophaga* spp. (Carrion et al. 2019), *Chryseobacterium* spp., *Flavobacterium* spp., *Microbacterium* spp., *Pseudomonas* spp., *Sphingomonas* spp. and *Stenotrophomonas* spp. (Kong et al. 2016; Liu et al. 2021) and *Xanthomonas* spp. (Berendsen et al. 2018). These microbes and their interactions with other plant-associated microbes and plant hosts might not only be beneficial when following a pathogen attack, but can also benefit future plant generations and reduce their stresses (Bakker et al. 2018; Berendsen et al. 2018). Based on the "cry for help" theory, a "DefenseBiome" concept has recently been proposed, which refers to the collective beneficial microbes induced by diseases in plants (Liu et al. 2020). The beneficial microbes within a "DefenseBiome" intimately interact with the plant immune system and/or inhibit plant pathogens and, in return, the plant immune system is modulated by the actions of a "DefenseBiome". A "DefenseBiome" that persists through plant microbiomes under different soil (e.g. types or physiochemical properties), plant (e.g. varieties or growth stages) and climate (e.g. temperature or rainfall) conditions may possess strong capacities to colonize plants, yielding significant options for application in agricultural production (Liu et al. 2020). Bioinoculants developed based on this strategy can possibly circumvent the ineffective colonization problem of traditional biocontrol strategies. A set of disease-induced metabolites secreted by plant roots to attract a "DefenseBiome" can also be used to induce the occurrence of beneficial microbes in soils.

9.4.2. *Microbial strategies for plant and soil microbiome engineering*

Those microorganisms systematically and consistently associating with a particular plant species across different environments are coined as "core" plant microbiomes (Hamonts et al. 2018). The core microbiomes in different crops, such as maize (Walters et al. 2018), rice (Edwards et al. 2015) and sugarcane (Hamonts et al. 2018) have been characterized and contain hundreds of co-occurring microbial taxa in each plant species. Hub microbes are the central component of a microbial co-occurrence network (Agler et al. 2016). They may possess abilities to recruit indigenous microorganisms with key biological and ecosystem functions; their presence or absence can prominently change the structure and functions of the plant/soil microbiomes (Toju et al. 2018). As such, core and hub microbes of the plant microbiome represent an appealing target to engineer microbiomes in situ in soil and use them to develop biopesticides or biostimulants that improve plant performance (Figure 9.2) (Qiu et al. 2019). There is evidence that the application of beneficial microbes on parental flowers leads to vertical transmission of the microbes across generations (Mitter et al. 2017). Such findings present a promising future for maintaining beneficial microbes via seed and flower microbiome engineering.

A recent study found that the combination of a sterilized biofertilizer with the biocontrol agent *Bacillus amyloliquefaciens* W19 led to the increase of a *Pseudomonas* spp. in the soil with biocontrol capacity against bananas' *Fusarium* wilt disease (Tao et al. 2020). This treatment had similar biocontrol effects as those of non-sterilized organic fertilizer. As such, biocontrol microbes could not only suppress a soil pathogen directly, but also increased plant disease tolerance via the altering of the soil microbial community and the enrichment of beneficial microbes.

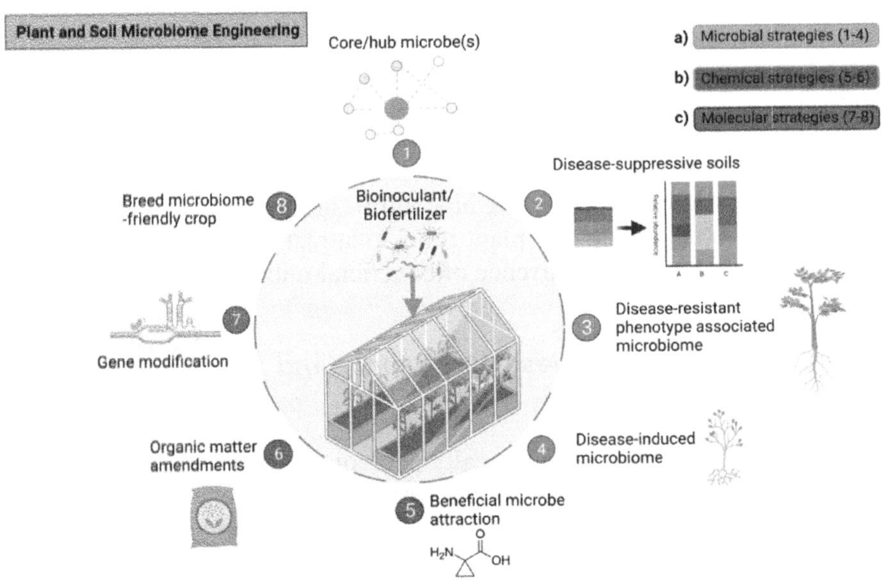

Figure 9.2. *Conceptual figure illustrating strategies for plant and soil microbiome engineering. This figure was created using Biorender.com. For a color version of this figure, see www.iste.co.uk/prigentcombaret/biocontrol.zip*

COMMENT ON FIGURE 9.2.– *At least three promising strategies can be wisely employed to engineer plant and soil microbiomes to enhance crop performance, including* ***(a)*** *microbial strategies (processes 1–4),* ***(b)*** *chemical strategies (processes 5 and 6) and* ***(c)*** *molecular strategies (processes 7 and 8). Plants have developed diverse strategies to strengthen their stress tolerance by modulating the composition, diversity and functions of their microbial symbionts (particularly those in the rhizosphere). As such, the microbes serving key host functions in nutrient acquisition and/or disease suppression can be used to develop bioinoculants to improve plant performance under stresses. These beneficial microbes (core/hub microbes) can be applied in soils, sprayed on plants or coated on seeds, which can consequently leverage plant and soil microbiomes towards fostering beneficial crop phenotypes (e.g. higher disease tolerance and nutrient utilization rates) (process 1). The key microbe(s) that suppress pathogens in disease suppressive soil/plant-associated microbiomes (e.g. DefenceBiome) are good candidates for bioinoculant/ biofertilizer development (processes 2–4). Root metabolites that attract beneficial microbes in the rhizosphere can be synthesized in large scale and applied in field to enrich specific beneficial microbes (process 5). In the same direction, soil quality, microbiome activity and plant health can be improved by applying organic amendments in the field (process 6). Finally, plant genes controlling a beneficial microbiome-mediated phenotype can also be transferred to other crops to boost their capacity in coping with particular harsh environments (e.g. drought/cold) (processes 7 and 8).*

Microbial transplantation is another way to engineer the soil microbiome and reduce disease incidence (Wei et al. 2019; Arif 2020). For example, transplantation of the *Flavobacterium*-enriched microbiome from the rhizosphere of a *R. solanacearum*-resistant tomato variety (Hawaii 7996) could suppress disease in susceptible plants (Kwak et al. 2018). Therefore, a native microbiome can play a key role in protecting plants from diseases. Current effective ways for delivering bioinoculants in soil include seed coating of microbes and the spraying of microbial suspensions on plants/soils, as well as the use of microbial capsules that gradually release microbes in soils (Qiu et al. 2019).

9.4.3. Chemical strategies for plant and soil microbiome engineering

Our knowledge of which plant metabolites attract which subset of microorganisms in the rhizosphere has been largely improved. The chemical molecules attracting beneficial microbial groups can be synthesized in large scale and applied in soil/plants to attract beneficial microbes and maintain soil microbial activities (Sasse et al. 2018; Pascale et al. 2020). 1-aminocyclopropane-1-carboxylate (ACC) is the immediate precursor of ethylene in plants, which is prominently involved in plant responses to diverse abiotic stresses such as drought, salinity and the cold (Duan et al. 2009). A recent study revealed that incubating saline soils with 2 mM ACC for two weeks improved ACC deaminase production and microbial properties in saline soil. ACC deaminases can prevent the overaccumulation of ACC in plants that can cause plant damages (Figure 9.2) (Liu et al. 2019a). As such, the beneficial effects generated in soils allowed lettuce plants to produce nearly five times more biomass than that in non-ACC treated soils. Overall, ACC has great potential in managing crop production in otherwise unusable saline soils (Arif et al. 2020).

In another study, already mentioned before, it was found that maize secretes flavones to enrich Oxalobacteraceae in the rhizosphere (Yu et al. 2021b). These microbes enhanced the growth and nitrogen acquisition of maize plants. This finding has applicable implications in agriculture given that it is possible to improve plant nitrogen acquisition by adding flavones in the soil to stimulate these beneficial microbes. Similarly, there are other root exudates that attract beneficial microbes and that increase plant stress resistance, including coumarins, citric acid, phenolic compounds and BXs (Rudrappa et al. 2008; Neal et al. 2012; Stringlis et al. 2018a; Kudjordjie et al. 2019; Harbort et al. 2020). A similar idea is to engineer seed microbiomes using particular chemicals to enhance plant stress resistance. For example, *Sphingomonas melonis* that is enriched and transmitted across generations in disease-resistant rice seeds can potentially produce anthranilic acid that likely inhibits the expression of virulence factors of the seed-borne pathogen *Burkholderia plantarii* (Matsumoto et al. 2021). However, it is worth noting that any safety concerns associated with the application of these chemicals must be fully addressed before going for large-scale application in field.

Volatile organic compounds (VOCs) represent a group of promising but less investigated chemicals that can be used to modulate the plant immune system and plant-associated microbiomes (Weisskopf et al. 2021). These compounds have the potential to manage crop health and growth in agriculture. For example, tomato plants inoculated with plant growth promoting bacterium *Bacillus amyloliquefaciens* strain GB03 released VOCs that triggered a salicylic acid signaling pathway of neighboring plants, and consequently allowed these plants to achieve microbial communities of similar structure in the rhizosphere with the VOC-emitting plants (Kong et al. 2020). However, the implications in plant health and growth are yet to be investigated. VOCs present only a small portion of root metabolites (<1%) but can directly interact with microorganisms in the rhizosphere and mediate intimate plant–microbe interactions. It was demonstrated that root VOCs of *Carex arenaria* could attract beneficial microbes from a remote distance (as far as 12 cm) to suppress the soilborne pathogen *Fusarium culmorum* (Schulz-Bohm et al. 2018). This study highlights a hidden role of plant VOCs in the mutualistic interactions between plants and microbes.

Plant hormones represent another group of plant-derived chemicals showing capacities in engineering plant and soil microbiomes (Trivedi et al. 2020). Previous studies revealed that salicylic acid and jasmonic acid-dependent signaling pathways could mediate changes in plant microbiomes in the rhizosphere (Carvalhais et al. 2013; Lebeis et al. 2015; Liu et al. 2017b). For example, application of methyl jasmonate (MeJA) on the aboveground plants could trigger changes in the rhizosphere microbiomes of wheat and *Arabidopsis* (Carvalhais et al. 2013). In the case of *Arabidopsis*, the microbes that are potentially associated with plant defense were enriched in the rhizosphere upon the activation of a JA signaling pathway (Carvalhais et al. 2013). This leads to the hypothesis that the application of plant hormones on plants is one of the options for engineering a plant microbiome towards enhanced disease tolerance (Liu et al. 2017a). However, potential detrimental effects of phytohormone application on plants should be addressed. For example, the application of high concentrations of these chemicals may cause significant inhibition on plant growth. Sensitivity of the plant responses to these hormones should also be considered. For example, the application of salicylic acid on wheat plant did not alter the wheat rhizosphere/root endosphere microbiome (Liu et al. 2018), which is in contrast with results in *Arabidopsis*, which showed that a salicylic acid signaling pathway is involved in the root microbiome assembly (Lebeis et al. 2015).

9.4.4. *Molecular strategies: genetic modification and breeding*

Plant genetics play an important role in determining host microbiome composition, abundance and function. Many tight associations exist between plant genomes and microbiomes (see section 9.3.1), which provides the basis to cultivate superior microbiome-mediated crop phenotypes via modifying plant genomes (Kroll et al. 2017). Molecular tools such as quantitative trait mapping can be used to identify genetic loci in plants that determine plant traits and plant genes controlling mutualistic plant–microbe interactions (Beilsmith et al. 2019; Deng et al. 2021) (Figure 9.2). The advent of the gene-editing tool CRISPR-Cas9 revolutionized the way of plant genome editing (Ran et al. 2013), and makes it possible to transfer genes involved in microbe–microbe and plant–microbe signaling pathways into plants, and manipulate the plant colonization by hub or core microbiome in a predictable fashion.

Plant–microbe interactions can also be optimized by breeding "microbe-friendly" crops that preferentially attract beneficial microbiome members to assist plants in nutrient acquisition and disease tolerance (Arif et al. 2020). As such, propagating a desired host phenotype (e.g. stress tolerance) can be achieved by allowing host plants to recruit certain beneficial microbes to their microbiomes across generations (Mueller and Sachs 2015). For example, microbiomes that influence the timing of flowering in *Arabidopsis* were selected via 10 generations' optimization based on the microbial effects on early- or late-flowering phenotype (Mitter et al. 2017). A similar approach can be used to breed disease- and stress-tolerant crops. However, it is worth noting that effects of environmental factors should be taken into consideration when applying the above-mentioned approaches for plant and soil microbiome engineering due to their extensive influences on plant–microbiome interactions.

9.5. Concluding remarks and future perspectives

Recent studies using metagenomics, metabolomics, metatranscriptomics and synthetic microbial communities have revealed diverse beneficial plant–microbe interactions and underlying molecular mechanisms. Chemical molecules derived from plants and microbes play an increasingly recognized role in shaping mutualistic interactions between plants and the beneficial microbial symbionts. Reduction of chemical pesticides and fertilizers used in agriculture, and by extension, relief of the environment from the toxic effects of those compounds, are goals that scientists have been pursuing for a long time. Employment of microbial, chemical and molecular tools for plant and soil microbiome engineering presents a promising future for the development of a sustainable agriculture. Our knowledge of plant–microbiome interactions and the applications of microbiome in modern agriculture are booming. However, our current capacity in engineering plant and soil microbiomes is still limited; more empirical and practical studies are urgently needed before we can widely use these methods in agriculture. In our vision, maximizing the potential of plant beneficial microbiomes will become a daily routine. We envision that the further development of this research field will pave the way for microbiome-assisted agriculture in the future.

9.6. References

Agler, M.T., Ruhe, J., Kroll, S., Morhenn, C., Kim, S.T., Weigel, D., Kemen, E.M. (2016). Microbial hub taxa link host and abiotic factors to plant microbiome variation. *PLoS Biology*, 14, e1002352.

Almario, J., Jeena, G., Wunder, J., Langen, G., Zuccaro, A., Coupland, G., Bucher, M. (2017). Root-associated fungal microbiota of nonmycorrhizal *Arabis alpina* and its contribution to plant phosphorus nutrition. *Proceedings of the National Academy of Sciences USA*, 114, 9403–9412.

Arif, I., Batool, M., Schenk, P.M. (2020). Plant microbiome engineering: Expected benefits for improved crop growth and resilience. *Trends in Biotechnology*, 38, 1385–1396.

Bai, Y., Muller, D.B., Srinivas, G., Garrido-Oter, R., Potthoff, E., Rott, M., Dombrowski, N., Munch, P.C., Spaepen, S., Remus-Emsermann, M. et al. (2015). Functional overlap of the *Arabidopsis* leaf and root microbiota. *Nature*, 528, 364–369.

Bakker, P.A.H.M., Pieterse, C.M.J., de Jonge, R., Berendsen, R.L. (2018). The soil-borne legacy. *Cell*, 172, 1178–1180.

Bakker, P.A.H.M., Berendsen, R.L., Van Pelt, J.A., Vismans, G., Yu, K., Li, E., Van Bentum, S., Poppeliers, S.W.M., Sanchez Gil, J.J., Zhang, H. et al. (2020). The soil-borne identity and microbiome-assisted agriculture: Looking back to the future. *Molecular Plant*, 13, 1394–1401.

Banerjee, S., Schlaeppi, K., Van der Heijden, M.G.V. (2018). Keystone taxa as drivers of microbiome structure and functioning. *Nature Reviews Microbiology*, 16, 567–576.

Beckers, B., De Beeck, M.O., Weyens, N., Van Acker, R., Van Montagu, M., Boerjan, W., Vangronsveld, J. (2016). Lignin engineering in field-grown poplar trees affects the endosphere bacterial microbiome. *Proceedings of the National Academy of Sciences USA*, 113, 2312–2317.

Beilsmith, K., Thoen, M.P.M., Brachi, B., Gloss, A.D., Khan, M.H., Bergelson, J. (2019). Genome-wide association studies on the phyllosphere microbiome: Embracing complexity in host-microbe interactions. *The Plant Journal*, 97, 164–181.

Beilsmith, K., Perisin, M., Bergelson, J. (2021). Natural bacterial assemblages in *Arabidopsis thaliana* tissues become more distinguishable and diverse during host development. *mBio*, 12, e02723-20.

Berendsen, R.L., Pieterse, C.M.J., Bakker, P.A.H.M. (2012). The rhizosphere microbiome and plant health. *Trends in Plant Science*, 17, 478–486.

Berendsen, R.L., Vismans, G., Yu, K., Song, Y., de Jonge, R., Burgman, W.P., Burmølle, M., Herschend, J., Bakker, P.A.H.M., Pieterse, C.M.J. (2018). Disease-induced assemblage of a plant-beneficial bacterial consortium. *The ISME Journal*, 12, 1496–1507.

Berg, G., Rybakova, D., Fischer, D., Cernava, T., Verges, M.C., Charles, T., Chen, X., Cocolin, L., Eversole, K., Corral, G.H. et al. (2020). Microbiome definition re-visited: Old concepts and new challenges. *Microbiome*, 8, 103.

Bergelson, J., Mittelstrass, J., Horton, M.W. (2019). Characterizing both bacteria and fungi improves understanding of the *Arabidopsis* root microbiome. *Scientific Reports*, 9, 24.

Boachon, B., Burdloff, Y., Ruan, J.-X., Rojo, R., Junker, R.R., Vincent, B., Nicolè, F., Bringel, F., Lesot, A., Henry, L. (2019). A promiscuous CYP706A3 reduces terpene volatile emission from *Arabidopsis* flowers, affecting florivores and the floral microbiome. *The Plant Cell*, 31, 2947–2972.

Bodenhausen, N., Bortfeld-Miller, M., Ackermann, M., Vorholt, J.A. (2014). A synthetic community approach reveals plant genotypes affecting the phyllosphere microbiota. *PLoS Genetics*, 10, e1004283.

Boyd, L.A., Ridout, C., O'Sullivan, D.M., Leach, J.E., Leung, H. (2013). Plant–pathogen interactions: Disease resistance in modern agriculture. *Trends in Genetics*, 29, 233–240.

Bressan, M., Roncato, M.A., Bellvert, F., Comte, G., Haichar, F.Z., Achouak, W., Berge, O. (2009). Exogenous glucosinolate produced by *Arabidopsis thaliana* has an impact on microbes in the rhizosphere and plant roots. *The ISME Journal*, 3, 1243–1257.

Bulgarelli, D., Rott, M., Schlaeppi, K., Ver Loren Van Themaat, E., Ahmadinejad, N., Assenza, F., Rauf, P., Huettel, B., Reinhardt, R., Schmelzer, E. et al. (2012). Revealing structure and assembly cues for *Arabidopsis* root-inhabiting bacterial microbiota. *Nature*, 488, 91–95.

Bulgarelli, D., Schlaeppi, K., Spaepen, S., Ver Loren Van Themaat, E., Schulze-Lefert, P. (2013). Structure and functions of the bacterial microbiota of plants. *Annual Reviews of Plant Biology*, 64, 807–838.

Bulgarelli, D., Garrido-Oter, R., Munch, P.C., Weiman, A., Droge, J., Pan, Y., McHardy, A.C., Schulze-Lefert, P. (2015). Structure and function of the bacterial root microbiota in wild and domesticated barley. *Cell Host & Microbe*, 17, 392–403.

Cadot, S., Guan, H., Bigalke, M., Walser, J.-C., Jander, G., Erb, M., Van der Heijden, M.G.A., Schlaeppi, K. (2021). Specific and conserved patterns of microbiota-structuring by maize benzoxazinoids in the field. *Microbiome*, 9, 103.

Carlstrom, C.I., Field, C.M., Bortfeld-Miller, M., Muller, B., Sunagawa, S., Vorholt, J.A. (2019). Synthetic microbiota reveal priority effects and keystone strains in the *Arabidopsis* phyllosphere. *Nature Ecology, Evolution*, 3, 1445–1454.

Carrion, V.J., Perez-Jaramillo, J., Cordovez, V., Tracanna, V., De Hollander, M., Ruiz-Buck, D., Mendes, L.W., Van Ijcken, W.F.J., Gomez-Exposito, R., Elsayed, S.S. et al. (2019). Pathogen-induced activation of disease-suppressive functions in the endophytic root microbiome. *Science*, 366, 606–612.

Carvalhais, L.C., Dennis, P.G., Badri, D.V., Tyson, G.W., Vivanco, J.M., Schenk, P.M. (2013). Activation of the jasmonic acid plant defence pathway alters the composition of rhizosphere bacterial communities. *PLoS One*, 8, e56457.

Carvalhais, L.C., Dennis, P.G., Badri, D.V., Kidd, B.N., Vivanco, J.M., Schenk, P.M. (2015). Linking jasmonic acid signaling, root exudates, and rhizosphere microbiomes. *Molecular Plant-Microbe Interactions*, 28, 1049–1058.

Carvalhais, L.C., Schenk, P.M., Dennis, P.G. (2017). Jasmonic acid signalling and the plant holobiont. *Current Opinion in Microbiology*, 37, 42–47.

Castrillo, G., Teixeira, P.J., Paredes, S.H., Law, T.F., De Lorenzo, L., Feltcher, M.E., Finkel, O.M., Breakfield, N.W., Mieczkowski, P., Jones, C.D. et al. (2017). Root microbiota drive direct integration of phosphate stress and immunity. *Nature*, 543, 513–518.

Chaparro, J.M., Badri, D.V., Bakker, M.G., Sugiyama, A., Manter, D.K., Vivanco, J.M. (2013). Root exudation of phytochemicals in *Arabidopsis* follows specific patterns that are developmentally programmed and correlate with soil microbial functions. *PLoS One*, 8, e55731.

Chaparro, J.M., Badri, D.V., Vivanco, J.M. (2014). Rhizosphere microbiome assemblage is affected by plant development. *The ISME Journal*, 8, 790–803.

Chapelle, E., Mendes, R., Bakker, P.A.H.M., Raaijmakers, J.M. (2016). Fungal invasion of the rhizosphere microbiome. *The ISME Journal*, 10, 265–268.

Cordovez, V., Dini-Andreote, F., Carrion, V.J., Raaijmakers, J.M. (2019). Ecology and evolution of plant microbiomes. *Annual Review of Microbiology*, 73, 69–88.

Cotton, T.E.A., Petriacq, P., Cameron, D.D., Meselmani, M.A., Schwarzenbacher, R., Rolfe, S.A., Ton, J. (2019). Metabolic regulation of the maize rhizobiome by benzoxazinoids. *The ISME Journal*, 13, 1647–1658.

Danhorn, T. and Fuqua, C. (2007). Biofilm formation by plant-associated bacteria. *Annual Review of Microbiology*, 61, 401–422.

De Souza, R.S., Okura, V.K., Armanhi, J.S., Jorrin, B., Lozano, N., Da Silva, M.J., Gonzalez-Guerrero, M., De Araujo, L.M., Verza, N.C., Bagheri, H.C. et al. (2016). Unlocking the bacterial and fungal communities assemblages of sugarcane microbiome. *Scientific Reports*, 6, 28774.

Dean, R., Van Kan, J.A., Pretorius, Z.A., Hammond-Kosack, K.E., Di Pietro, A., Spanu, P.D., Rudd, J.J., Dickman, M., Kahmann, R., Ellis, J. et al. (2012). The top 10 fungal pathogens in molecular plant pathology. *Molecular Plant Pathology*, 13, 414–430.

Deng, S., Caddell, D.F., Xu, G., Dahlen, L., Washington, L., Yang, J., Coleman-Derr, D. (2021). Genome wide association study reveals plant loci controlling heritability of the rhizosphere microbiome. *The ISME Journal*, 15, 3181–3194.

Dombrowski, N., Schlaeppi, K., Agler, M.T., Hacquard, S., Kemen, E., Garrido-Oter, R., Wunder, J., Coupland, G., Schulze-Lefert, P. (2017). Root microbiota dynamics of perennial *Arabis alpina* are dependent on soil residence time but independent of flowering time. *The ISME Journal*, 11, 43–55.

Duan, J., Muller, K.M., Charles, T.C., Vesely, S., Glick, B.R. (2009). 1-aminocyclopropane-1-carboxylate (ACC) deaminase genes in rhizobia from southern Saskatchewan. *Microbial Ecology*, 57, 423–436.

Duran, P., Thiergart, T., Garrido-Oter, R., Agler, M., Kemen, E., Schulze-Lefert, P., Hacquard, S. (2018). Microbial interkingdom interactions in roots promote *Arabidopsis* survival. *Cell*, 175, 973–983.

Durr, J., Reyt, G., Spaepen, S., Hilton, S., Meehan, C., Qi, W., Kamiya, T., Flis, P., Dickinson, H.G., Feher, A. (2021). A novel signaling pathway required for *Arabidopsis* endodermal root organization shapes the rhizosphere microbiome. *Plant and Cell Physiology*, 62, 248–261.

Edwards, J., Johnson, C., Santos-Medellin, C., Lurie, E., Podishetty, N.K., Bhatnagar, S., Eisen, J.A., Sundaresan, V. (2015). Structure, variation, and assembly of the root-associated microbiomes of rice. *Proceedings of the National Academy of Sciences USA*, 112, 911–920.

Fang, F.C. (2011). Antimicrobial actions of reactive oxygen species. *mBio*, 2, e00141-11.

Finkel, O.M., Castrillo, G., Herrera Paredes, S., Salas-González, I., Dangl, J.L. (2017). Understanding and exploiting plant beneficial microbes. *Current Opinion in Plant Biology*, 38, 155–163.

Finkel, O.M., Salas-González, I., Castrillo, G., Conway, J.M., Law, T.F., Teixeira, P.J.P.L., Wilson, E.D., Fitzpatrick, C.R., Jones, C.D., Dangl, J.L. (2020). A single bacterial genus maintains root growth in a complex microbiome. *Nature*, 587, 103–108.

Fitzpatrick, C.R., Copeland, J., Wang, P.W., Guttman, D.S., Kotanen, P.M., Johnson, M.T.J. (2018). Assembly and ecological function of the root microbiome across angiosperm plant species. *Proceedings of the National Academy of Sciences USA*, 115, E1157–E1165.

Gage, D.J. (2004). Infection and invasion of roots by symbiotic, nitrogen-fixing rhizobia during nodulation of temperate legumes. *Microbiology and Molecular Biology Reviews*, 68, 280–300.

Gao, Z., Karlsson, I., Geisen, S., Kowalchuk, G., Jousset, A. (2019). Protists: Puppet masters of the rhizosphere microbiome. *Trends in Plant Science*, 24, 165–176.

Genre, A., Chabaud, M., Faccio, A., Barker, D.G., Bonfante, P. (2008). Prepenetration apparatus assembly precedes and predicts the colonization patterns of arbuscular mycorrhizal fungi within the root cortex of both *Medicago truncatula* and *Daucus carota*. *The Plant Cell*, 20, 1407–1420.

Genre, A., Lanfranco, L., Perotto, S., Bonfante, P. (2020). Unique and common traits in mycorrhizal symbioses. *Nature Reviews Microbiology*, 18, 649–660.

Glawischnig, E. (2007). Camalexin. *Phytochemistry*, 68, 401–406.

Goldford, J.E., Lu, N., Bajić, D., Estrela, S., Tikhonov, M., Sanchez-Gorostiaga, A., Segrè, D., Mehta, P., Sanchez, A. (2018). Emergent simplicity in microbial community assembly. *Science*, 361, 469–474.

Gomiero, T., Pimentel, D., Paoletti, M.G. (2011). Is there a need for a more sustainable agriculture? *Critical Reviews in Plant Sciences*, 30, 6–23.

Gray, A.N., Koo, B.-M., Shiver, A.L., Peters, J.M., Osadnik, H., Gross, C.A. (2015). High-throughput bacterial functional genomics in the sequencing era. *Current Opinion in Microbiology*, 27, 86–95.

Gu, S., Wei, Z., Shao, Z., Friman, V.P., Cao, K., Yang, T., Kramer, J., Wang, X., Li, M., Mei, X. et al. (2020). Competition for iron drives phytopathogen control by natural rhizosphere microbiomes. *Nature Microbiology*, 5, 1002–1010.

Hacquard, S., Spaepen, S., Garrido-Oter, R., Schulze-Lefert, P. (2017). Interplay between innate immunity and the plant microbiota. *Annual Review of Phytopathology*, 55, 565–589.

Hamonts, K., Trivedi, P., Garg, A., Janitz, C., Grinyer, J., Holford, P., Botha, F.C., Anderson, I.C., Singh, B.K. (2018). Field study reveals core plant microbiota and relative importance of their drivers. *Environmental Microbiology*, 20, 124–140.

Haney, C.H., Wiesmann, C.L., Shapiro, L.R., Melnyk, R.A., O'Sullivan, L.R., Khorasani, S., Xiao, L., Han, J., Bush, J., Carrillo, J. et al. (2017). Rhizosphere-associated *Pseudomonas* induce systemic resistance to herbivores at the cost of susceptibility to bacterial pathogens. *Molecular Ecology*, 27, 1833–1847.

Harbort, C.J., Hashimoto, M., Inoue, H., Niu, Y., Guan, R., Rombolà, A.D., Kopriva, S., Voges, M.J.E.E.E., Sattely, E.S., Garrido-Oter, R. et al. (2020). Root-secreted coumarins and the microbiota interact to improve iron nutrition in *Arabidopsis*. *Cell Host & Microbe*, 28, 825–837.

Hartmann, A., Rothballer, M., Schmid, M. (2008). Lorenz Hiltner, a pioneer in rhizosphere microbial ecology and soil bacteriology research. *Plant and Soil*, 312, 7–14.

Hassani, M.A., Duran, P., Hacquard, S. (2018). Microbial interactions within the plant holobiont. *Microbiome*, 6, 58.

Hiruma, K., Gerlach, N., Sacristan, S., Nakano, R.T., Hacquard, S., Kracher, B., Neumann, U., Ramirez, D., Bucher, M., O'Connell, R.J. et al. (2016). Root endophyte *Colletotrichum tofieldiae* confers plant fitness benefits that are phosphate status dependent. *Cell*, 165, 464–474.

Hou, S., Thiergart, T., Vannier, N., Mesny, F., Ziegler, J., Pickel, B., Hacquard, S. (2020). A microbiota-root-shoot circuit favours *Arabidopsis* growth over defence under suboptimal light. *Nature Plants*, 7, 1078–1092.

Hu, L., Robert, C.A.M., Cadot, S., Zhang, X., Ye, M., Li, B., Manzo, D., Chervet, N., Steinger, T., Van der Heijden, M.G.A. et al. (2018). Root exudate metabolites drive plant-soil feedbacks on growth and defense by shaping the rhizosphere microbiota. *Nature Communications*, 9, 2738.

Huang, A.C., Jiang, T., Liu, Y.X., Bai, Y.C., Reed, J., Qu, B., Goossens, A., Nutzmann, H.W., Bai, Y., Osbourn, A. (2019). A specialized metabolic network selectively modulates *Arabidopsis* root microbiota. *Science*, 364, eaau6389.

Huot, B., Yao, J., Montgomery, B.L., He, S.Y. (2014). Growth-defense tradeoffs in plants: A balancing act to optimize fitness. *Molecular Plant*, 7, 1267–1287.

Jiao, W., Chen, W., Chang, A.C., Page, A.L. (2012). Environmental risks of trace elements associated with long-term phosphate fertilizers applications: A review. *Environmental Pollution*, 168, 44–53.

Jung, J., Kim, J.S., Taffner, J., Berg, G., Ryu, C.M. (2020). Archaea, tiny helpers of land plants. *Computational and Structural Biotechnology Journal*, 18, 2494–2500.

Kamoun, S., Furzer, O., Jones, J.D., Judelson, H.S., Ali, G.S., Dalio, R.J., Roy, S.G., Schena, L., Zambounis, A., Panabieres, F. et al. (2015). The top 10 oomycete pathogens in molecular plant pathology. *Molecular Plant Pathology*, 16, 413–434.

Kong, H.G., Kim, B.K., Song, G.C., Lee, S., Ryu, C.M. (2016). Aboveground whitefly infestation-mediated reshaping of the root microbiota. *Frontiers in Microbiology*, 7, 1314.

Kong, H.G., Song, G.C., Sim, H.J., Ryu, C.M. (2020). Achieving similar root microbiota composition in neighbouring plants through airborne signalling. *The ISME Journal*, 15, 397–408.

Koprivova, A., Schuck, S., Jacoby, R.P., Klinkhammer, I., Welter, B., Leson, L., Martyn, A., Nauen, J., Grabenhorst, N., Mandelkow, J.F. et al. (2019). Root-specific camalexin biosynthesis controls the plant growth-promoting effects of multiple bacterial strains. *Proceedings of the National Academy of Sciences USA*, 116, 15735–15744.

Kroll, S., Agler, M.T., Kemen, E. (2017). Genomic dissection of host-microbe and microbe-microbe interactions for advanced plant breeding. *Current Opinion in Plant Biology*, 36, 71–78.

Kudjordjie, E.N., Sapkota, R., Steffensen, S.K., Fomsgaard, I.S., Nicolaisen, M. (2019). Maize synthesized benzoxazinoids affect the host associated microbiome. *Microbiome*, 7, 59.

Kwak, M.J., Kong, H.G., Choi, K., Kwon, S.K., Song, J.Y., Lee, J., Lee, P.A., Choi, S.Y., Seo, M., Lee, H.J. et al. (2018). Rhizosphere microbiome structure alters to enable wilt resistance in tomato. *Nature Biotechnology*, 36, 1100–1109.

Lebeis, S.L., Paredes, S.H., Lundberg, D.S., Breakfield, N., Gehring, J., McDonald, M., Malfatti, S., Glavina del Rio, T., Jones, C.D., Tringe, S.G. et al. (2015). Salicylic acid modulates colonization of the root microbiome by specific bacterial taxa. *Science*, 349, 860–864.

Lee, S.M. and Ryu, C.M. (2021). Algae as new kids in the beneficial plant microbiome. *Frontiers in Plant Science*, 12, 599742.

Levy, A., Salas Gonzalez, I., Mittelviefhaus, M., Clingenpeel, S., Herrera Paredes, S., Miao, J., Wang, K., Devescovi, G., Stillman, K., Monteiro, F. et al. (2018). Genomic features of bacterial adaptation to plants. *Nature Genetics*, 50, 138–150.

Liu, H., Carvalhais, L.C., Crawford, M., Singh, E., Dennis, P.G., Pieterse, C.M.J., Schenk, P.M. (2017a). Inner plant values: Diversity, colonization and benefits from endophytic bacteria. *Frontiers in Microbiology*, 8, 2552.

Liu, H., Carvalhais, L.C., Schenk, P.M., Dennis, P.G. (2017b). Effects of jasmonic acid signalling on the wheat microbiome differ between body sites. *Scientific Reports*, 7, 41766.

Liu, H., Carvalhais, L.C., Schenk, P.M., Dennis, P.G. (2018). Activation of the salicylic acid signalling pathway in wheat had no significant short-term impact on the diversity of root-associated microbiomes. *Pedobiologia*, 70, 6–11.

Liu, H., Khan, M.Y., Carvalhais, L.C., Delgado-Baquerizo, M., Yan, L., Crawford, M., Dennis, P.G., Singh, B., Schenk, P.M. (2019a). Soil amendments with ethylene precursor alleviate negative impacts of salinity on soil microbial properties and productivity. *Scientific Reports*, 9, 6892.

Liu, Y., Zhu, A., Tan, H., Cao, L., Zhang, R. (2019b). Engineering banana endosphere microbiome to improve *Fusarium* wilt resistance in banana. *Microbiome*, 7, 74.

Liu, H., Brettell, L.E., Qiu, Z., Singh, B.K. (2020). Microbiome-mediated stress resistance in plants. *Trends in Plant Science*, 25, 733–743.

Liu, H., Li, J., Carvalhais, L.C., Percy, C.D., Prakash Verma, J., Schenk, P.M., Singh, B.K. (2021). Evidence for the plant recruitment of beneficial microbes to suppress soil-borne pathogens. *New Phytologist*, 229, 2873–2885.

Lugtenberg, B. and Kamilova, F. (2009). Plant-growth-promoting rhizobacteria. *Annual Review of Microbiology*, 63, 541–556.

Luna, E., Pastor, V., Robert, J., Flors, V., Mauch-Mani, B., Ton, J. (2011). Callose deposition: A multifaceted plant defense response. *Molecular Plant-Microbe Interactions*, 24, 183–193.

Lundberg, D.S., Lebeis, S.L., Paredes, S.H., Yourstone, S., Gehring, J., Malfatti, S., Tremblay, J., Engelbrektson, A., Kunin, V., Del Rio, T.G. et al. (2012). Defining the core *Arabidopsis thaliana* root microbiome. *Nature*, 488, 86–90.

Ma, K.W., Niu, Y., Jia, Y., Ordon, J., Copeland, C., Emonet, A., Geldner, N., Guan, R., Stolze, S.C., Nakagami, H. et al. (2021). Coordination of microbe–host homeostasis by crosstalk with plant innate immunity. *Nature Plants*, 7, 814–825.

Mann, R.M., Hyne, R.V., Choung, C.B., Wilson, S.P. (2009). Amphibians and agricultural chemicals: Review of the risks in a complex environment. *Environmental Pollution*, 157, 2903–2927.

Mansfield, J., Genin, S., Magori, S., Citovsky, V., Sriariyanum, M., Ronald, P., Dow, M., Verdier, V., Beer, S.V., Machado, M.A. et al. (2012). Top 10 plant pathogenic bacteria in molecular plant pathology. *Molecular Plant Pathology*, 13, 614–629.

Matsumoto, H., Fan, X., Wang, Y., Kusstatscher, P., Duan, J., Wu, S., Chen, S., Qiao, K., Wang, Y., Ma, B. et al. (2021). Bacterial seed endophyte shapes disease resistance in rice. *Nature Plants*, 7, 60–72.

Mazurier, S., Corberand, T., Lemanceau, P., Raaijmakers, J.M. (2009). Phenazine antibiotics produced by fluorescent pseudomonads contribute to natural soil suppressiveness to *Fusarium* wilt. *The ISME Journal*, 3, 977–991.

Mendes, R., Kruijt, M., de Bruijn, I., Dekkers, E., van der Voort, M., Schneider, J.H., Piceno, Y.M., Desantis, T.Z., Andersen, G.L., Bakker, P.A.H.M. et al. (2011). Deciphering the rhizosphere microbiome for disease-suppressive bacteria. *Science*, 332, 1097–1100.

Mendes, L.W., Raaijmakers, J.M., de Hollander, M., Mendes, R., Tsai, S.M. (2018). Influence of resistance breeding in common bean on rhizosphere microbiome composition and function. *The ISME Journal*, 12, 212–224.

Millet, Y.A., Danna, C.H., Clay, N.K., Songnuan, W., Simon, M.D., Werck-Reichhart, D., Ausubel, F.M. (2010). Innate immune responses activated in *Arabidopsis* roots by microbe-associated molecular patterns. *Plant Cell*, 22, 973–990.

Mitter, B., Pfaffenbichler, N., Flavell, R., Compant, S., Antonielli, L., Petric, A., Berninger, T., Naveed, M., Sheibani-Tezerji, R., von Maltzahn, G. et al. (2017). A new approach to modify plant microbiomes and traits by introducing beneficial bacteria at flowering into progeny seeds. *Frontiers in Microbiology*, 8, 11.

Mittler, R. (2006). Abiotic stress, the field environment and stress combination. *Trends in Plant Science*, 11, 15–19.

Mittler, R. (2017). ROS are good. *Trends in Plant Science*, 22, 11–19.

Mueller, U.G. and Sachs, J.L. (2015). Engineering microbiomes to improve plant and animal health. *Trends in Microbiology*, 23, 606–617.

Neal, A.L., Ahmad, S., Gordon-Weeks, R., Ton, J. (2012). Benzoxazinoids in root exudates of maize attract *Pseudomonas putida* to the rhizosphere. *PLoS One*, 7, e35498.

Ngou, B.P.M., Ahn, H.K., Ding, P., Jones, J.D.G. (2021). Mutual potentiation of plant immunity by cell-surface and intracellular receptors. *Nature*, 592, 110–115.

Nicolopoulou-Stamati, P., Maipas, S., Kotampasi, C., Stamatis, P., Hens, L. (2016). Chemical pesticides and human health: The urgent need for a new concept in agriculture. *Frontiers in Public Health*, 4, 148.

Oldroyd, G.E., Murray, J.D., Poole, P.S., Downie, J.A. (2011). The rules of engagement in the legume-rhizobial symbiosis. *Annual Review of Genetics*, 45, 119–144.

Pascale, A., Proietti, S., Pantelides, I.S., Stringlis, I.A. (2020). Modulation of the root microbiome by plant molecules: The basis for targeted disease suppression and plant growth promotion. *Frontiers in Plant Science*, 10, 1741.

Peiffer, J.A., Spor, A., Koren, O., Jin, Z., Tringe, S.G., Dangl, J.L., Buckler, E.S., Ley, R.E. (2013). Diversity and heritability of the maize rhizosphere microbiome under field conditions. *Proceedings of the National Academy of Sciences USA*, 110, 6548–6553.

Pérez-Jaramillo, J.E., de Hollander, M., Ramírez, C.A., Mendes, R., Raaijmakers, J.M., Carrión, V.J. (2019). Deciphering rhizosphere microbiome assembly of wild and modern common bean (*Phaseolus vulgaris*) in native and agricultural soils from Colombia. *Microbiome*, 7, 114.

Pfeilmeier, S., Petti, G.C., Bortfeld-Miller, M., Daniel, B., Field, C.M., Sunagawa, S., Vorholt, J.A. (2021). The plant NADPH oxidase RBOHD is required for microbiota homeostasis in leaves. *Nature Microbiology*, 6, 852–864.

Piasecka, A., Jedrzejczak-Rey, N., Bednarek, P. (2015). Secondary metabolites in plant innate immunity: Conserved function of divergent chemicals. *New Phytologist*, 206, 948–964.

Pieterse, C.M.J., Zamioudis, C., Berendsen, R.L., Weller, D.M., Van Wees, S.C.M., Bakker, P.A.H.M. (2014). Induced systemic resistance by beneficial microbes. *Annual Review of Phytopathology*, 52, 347–375.

Qiu, Z., Egidi, E., Liu, H., Kaur, S., Singh, B.K. (2019). New frontiers in agriculture productivity: Optimised microbial inoculants and in situ microbiome engineering. *Biotechnology Advances*, 37, 107371.

Ran, F.A., Hsu, P.D., Wright, J., Agarwala, V., Scott, D.A., Zhang, F. (2013). Genome engineering using the CRISPR-Cas9 system. *Nature Protocols*, 8, 2281–2308.

Raza, W., Wang, J., Jousset, A., Friman, V.P., Mei, X., Wang, S., Wei, Z., Shen, Q. (2020). Bacterial community richness shifts the balance between volatile organic compound-mediated microbe–pathogen and microbe–plant interactions. *Proceedings of the Royal Society B: Biological Sciences*, 287, 20200403.

Ritpitakphong, U., Falquet, L., Vimoltust, A., Berger, A., Metraux, J.P., L'Haridon, F. (2016). The microbiome of the leaf surface of *Arabidopsis* protects against a fungal pathogen. *New Phytologist*, 210, 1033–1043.

Robertson-Albertyn, S., Alegria Terrazas, R., Balbirnie, K., Blank, M., Janiak, A., Szarejko, I., Chmielewska, B., Karcz, J., Morris, J., Hedley, P.E. et al. (2017). Root hair mutations displace the barley rhizosphere microbiota. *Frontiers in Plant Science*, 8, 1094.

Rolli, E., Marasco, R., Vigani, G., Ettoumi, B., Mapelli, F., Deangelis, M.L., Gandolfi, C., Casati, E., Previtali, F., Gerbino, R. et al. (2015). Improved plant resistance to drought is promoted by the root-associated microbiome as a water stress-dependent trait. *Environmental Microbiology*, 17, 316–331.

Roppolo, D., Boeckmann, B., Pfister, A., Boutet, E., Rubio, M.C., Denervaud-Tendon, V., Vermeer, J.E., Gheyselinck, J., Xenarios, I., Geldner, N. (2014). Functional and evolutionary analysis of the CASPARIAN STRIP MEMBRANE DOMAIN PROTEIN family. *Plant Physiology*, 165, 1709–1722.

Roser, M. (2013). Future population growth. *Our World in Data* [Online]. Available at: https://ourworldindata.org/future-population-growth.

Rudrappa, T., Czymmek, K.J., Pare, P.W., Bais, H.P. (2008). Root-secreted malic acid recruits beneficial soil bacteria. *Plant Physiology*, 148, 1547–1556.

Salas-Gonzalez, I., Reyt, G., Flis, P., Custodio, V., Gopaulchan, D., Bakhoum, N., Dew, T.P., Suresh, K., Franke, R.B., Dangl, J.L. et al. (2021). Coordination between microbiota and root endodermis supports plant mineral nutrient homeostasis. *Science*, 371, eabd0695.

Sapp, M., Ploch, S., Fiore-Donno, A.M., Bonkowski, M., Rose, L.E. (2018). Protists are an integral part of the *Arabidopsis thaliana* microbiome. *Environmental Microbiology*, 20, 30–43.

Sasse, J., Martinoia, E., Northen, T. (2018). Feed your friends: Do plant exudates shape the root microbiome? *Trends in Plant Science*, 23, 25–41.

Schmid, N.B., Giehl, R.F., Doll, S., Mock, H.P., Strehmel, N., Scheel, D., Kong, X., Hider, R.C., Von Wiren, N. (2014). Feruloyl-CoA 6'-Hydroxylase1-dependent coumarins mediate iron acquisition from alkaline substrates in *Arabidopsis*. *Plant Physiology*, 164, 160–172.

Schreiber, L. (2010). Transport barriers made of cutin, suberin and associated waxes. *Trends in Plant Science*, 15, 546–553.

Schulz-Bohm, K., Gerards, S., Hundscheid, M., Melenhorst, J., De Boer, W., Garbeva, P. (2018). Calling from distance: Attraction of soil bacteria by plant root volatiles. *The ISME Journal*, 12, 1252–1262.

Sczyrba, A., Hofmann, P., Belmann, P., Koslicki, D., Janssen, S., Droge, J., Gregor, I., Majda, S., Fiedler, J., Dahms, E. et al. (2017). Critical assessment of metagenome interpretation – A benchmark of metagenomics software. *Nature Methods*, 14, 1063–1071.

Serrano, M., Coluccia, F., Torres, M., L'Haridon, F., Metraux, J.P. (2014). The cuticle and plant defense to pathogens. *Frontiers in Plant Science*, 5, 274.

Snelders, N.C., Rovenich, H., Petti, G.C., Rocafort, M., Van Den Berg, G.C.M., Vorholt, J.A., Mesters, J.R., Seidl, M.F., Nijland, R., Thomma, B. (2020). Microbiome manipulation by a soil-borne fungal plant pathogen using effector proteins. *Nature Plants*, 6, 1365–1374.

Song, Y., Wilson, A.J., Zhang, X.C., Thoms, D., Sohrabi, R., Song, S., Geissmann, Q., Liu, Y., Walgren, L., He, S.Y. et al. (2021). FERONIA restricts *Pseudomonas* in the rhizosphere microbiome via regulation of reactive oxygen species. *Nature Plants*, 7, 644–654.

Stassen, M.J.J., Hsu, S.-H., Pieterse, C.M.J., Stringlis, I.A. (2021). Coumarin communication along the microbiome–root–shoot axis. *Trends in Plant Science*, 26, 169–183.

Stringlis, I.A. and Pieterse, C.M.J. (2021). Evolutionary "hide and seek" between bacterial flagellin and the plant immune system. *Cell Host & Microbe*, 29, 548–550.

Stringlis, I.A., Yu, K., Feussner, K., de Jonge, R., Van Bentum, S., Van Verk, M.C., Berendsen, R.L., Bakker, P.A.H.M., Feussner, I., Pieterse, C.M.J. (2018a). MYB72-dependent coumarin exudation shapes root microbiome assembly to promote plant health. *Proceedings of the National Academy of Sciences USA*, 115, E5213–E5222.

Stringlis, I.A., Zhang, H., Pieterse, C.M.J., Bolton, M.D., de Jonge, R. (2018b) Microbial small molecules – Weapons of plant subversion. *Natural Product Reports*, 35, 410–433.

Stringlis, I.A., de Jonge, R., Pieterse, C.M.J. (2019). The age of coumarins in plant–microbe interactions. *Plant Cell and Physiology*, 60, 1405–1419.

Taffner, J., Erlacher, A., Bragina, A., Berg, C., Moissl-Eichinger, C., Berg, G. (2018). What is the role of Archaea in plants? New insights from the vegetation of Alpine bogs. *mSphere*, 3, e00122-18.

Tao, C., Li, R., Xiong, W., Shen, Z., Liu, S., Wang, B., Ruan, Y., Geisen, S., Shen, Q., Kowalchuk, G.A. (2020). Bio-organic fertilizers stimulate indigenous soil *Pseudomonas* populations to enhance plant disease suppression. *Microbiome*, 8, 1–14.

Teixeira, P.J.P., Colaianni, N.R., Fitzpatrick, C.R., Dangl, J.L. (2019). Beyond pathogens: Microbiota interactions with the plant immune system. *Current Opinion in Microbiology*, 49, 7–17.

Teixeira, P.J.P., Colaianni, N.R., Law, T.F., Conway, J.M., Gilbert, S., Li, H., Salas-Gonzalez, I., Panda, D., Del Risco, N.M., Finkel, O.M. et al. (2021). Specific modulation of the root immune system by a community of commensal bacteria. *Proceedings of the National Academy of Sciences USA*, 118, e2100678118.

Thimmappa, R., Geisler, K., Louveau, T., O'Maille, P., Osbourn, A. (2014). Triterpene biosynthesis in plants. *Annual Review of Plant Biology*, 65, 225–257.

Tobimatsu, Y. and Schuetz, M. (2019). Lignin polymerization: How do plants manage the chemistry so well? *Current Opinion in Biotechnology*, 56, 75–81.

Toju, H., Peay, K.G., Yamamichi, M., Narisawa, K., Hiruma, K., Naito, K., Fukuda, S., Ushio, M., Nakaoka, S., Onoda, Y. (2018). Core microbiomes for sustainable agroecosystems. *Nature Plants*, 4, 247–257.

Trivedi, P., Leach, J.E., Tringe, S.G., Sa, T., Singh, B.K. (2020). Plant-microbiome interactions: From community assembly to plant health. *Nature Reviews Microbiology*, 18, 607–621.

Tsai, H.H. and Schmidt, W. (2017). Mobilization of iron by plant-borne coumarins. *Trends in Plant Science*, 22, 538–548.

Van Deynze, A., Zamora, P., Delaux, P.-M., Heitmann, C., Jayaraman, D., Rajasekar, S., Graham, D., Maeda, J., Gibson, D.M., Schwartz, K.D. (2018). Nitrogen fixation in a landrace of maize is supported by a mucilage-associated diazotrophic microbiota. *PLoS Biology*, 16, e2006352.

Voges, M., Bai, Y., Schulze-Lefert, P., Sattely, E.S. (2019). Plant-derived coumarins shape the composition of an *Arabidopsis* synthetic root microbiome. *Proceedings of the National Academy of Sciences USA*, 116, 12558–12565.

Walters, W.A., Jin, Z., Youngblut, N., Wallace, J.G., Sutter, J., Zhang, W., González-Peña, A., Peiffer, J.A., Koren, O., Shi, Q. (2018). Large-scale replicated field study of maize rhizosphere identifies heritable microbes. *Proceedings of the National Academy of Sciences USA*, 115, 7368–7373.

Wang, S., Alseekh, S., Fernie, A.R., Luo, J. (2019a). The structure and function of major plant metabolite modifications. *Molecular Plant*, 12, 899–919.

Wang, X., Wei, Z., Yang, K., Wang, J., Jousset, A., Xu, Y., Shen, Q., Friman, V.P. (2019b). Phage combination therapies for bacterial wilt disease in tomato. *Nature Biotechnology*, 37, 1513–1520.

Wei, Z. and Jousset, A. (2017). Plant breeding goes microbial. *Trends in Plant Science*, 22, 555–558.

Wei, Z., Gu, Y., Friman, V.-P., Kowalchuk, G.A., Xu, Y., Shen, Q., Jousset, A. (2019). Initial soil microbiome composition and functioning predetermine future plant health. *Science Advances*, 5, eaaw0759.

Weisskopf, L., Schulz, S., Garbeva, P. (2021). Microbial volatile organic compounds in intra-kingdom and inter-kingdom interactions. *Nature Reviews Microbiology*, 19, 391–404.

Weller, D.M., Raaijmakers, J.M., Gardener, B.B., Thomashow, L.S. (2002). Microbial populations responsible for specific soil suppressiveness to plant pathogens. *Annual Review of Phytopathology*, 40, 309–348.

Xiong, W., Song, Y., Yang, K., Gu, Y., Wei, Z., Kowalchuk, G.A., Xu, Y., Jousset, A., Shen, Q., Geisen, S. (2020). Rhizosphere protists are key determinants of plant health. *Microbiome*, 8, 27.

Xu, L., Naylor, D., Dong, Z., Simmons, T., Pierroz, G., Hixson, K.K., Kim, Y.M., Zink, E.M., Engbrecht, K.M., Wang, Y. et al. (2018). Drought delays development of the sorghum root microbiome and enriches for monoderm bacteria. *Proceedings of the National Academy of Sciences USA*, 115, E4284–E4293.

Yu, K., Pieterse, C.M.J., Bakker, P.A.H.M., Berendsen, R.L. (2019). Beneficial microbes going underground of root immunity. *Plant Cell and Environment*, 42, 2860–2870.

Yu, K., Stringlis, I.A., Van Bentum, S., de Jonge, R., Snoek, B.L., Pieterse, C.M.J., Bakker, P.A.H.M., Berendsen, R.L. (2021a). Transcriptome signatures in *Pseudomonas simiae* WCS417 shed light on role of root-secreted coumarins in *Arabidopsis*-mutualist communication. *Microorganisms*, 9, 575.

Yu, P., He, X., Baer, M., Beirinckx, S., Tian, T., Moya, Y.A.T., Zhang, X., Deichmann, M., Frey, F.P., Bresgen, V. (2021b). Plant flavones enrich rhizosphere *Oxalobacteraceae* to improve maize performance under nitrogen deprivation. *Nature Plants*, 7, 481–499.

Yuan, J., Zhao, J., Wen, T., Zhao, M., Li, R., Goossens, P., Huang, Q., Bai, Y., Vivanco, J.M., Kowalchuk, G.A. et al. (2018). Root exudates drive the soil-borne legacy of aboveground pathogen infection. *Microbiome*, 6, 156.

Yuan, M., Jiang, Z., Bi, G., Nomura, K., Liu, M., Wang, Y., Cai, B., Zhou, J.M., He, S.Y., Xin, X.F. (2021). Pattern-recognition receptors are required for NLR-mediated plant immunity. *Nature*, 592, 105–109.

Zamioudis, C., Korteland, J., Van Pelt, J.A., Van Hamersveld, M., Dombrowski, N., Bai, Y., Hanson, J., Van Verk, M.C., Ling, H.Q., Schulze-Lefert, P. et al. (2015). Rhizobacterial volatiles and photosynthesis-related signals coordinate *MYB72* expression in *Arabidopsis* roots during onset of induced systemic resistance and iron-deficiency responses. *Plant Journal*, 84, 309–322.

Zhang, Y., Xu, J., Riera, N., Jin, T., Li, J., Wang, N. (2017). Huanglongbing impairs the rhizosphere-to-rhizoplane enrichment process of the citrus root-associated microbiome. *Microbiome*, 5, 97.

Zhang, J.Y., Liu, Y.X., Zhang, N., Hu, B., Jin, T., Xu, H.R., Qin, Y., Yan, P.X., Zhang, X.N., Guo, X.X. et al. (2019). NRT1.1B is associated with root microbiota composition and nitrogen use in field-grown rice. *Nature Biotechnology*, 37, 676–684.

Zhang, J., Liu, Y.X., Guo, X., Qin, Y., Garrido-Oter, R., Schulze-Lefert, P., Bai, Y. (2021). High-throughput cultivation and identification of bacteria from the plant root microbiota. *Nature Protocols*, 16, 988–1012.

Zhou, S., Richter, A., Jander, G. (2018). Beyond defense: Multiple functions of benzoxazinoids in maize metabolism. *Plant Cell Physiology*, 59, 1528–1537.

Zhou, F., Emonet, A., Denervaud Tendon, V., Marhavy, P., Wu, D., Lahaye, T., Geldner, N. (2020). Co-incidence of damage and microbial patterns controls localized immune responses in roots. *Cell*, 180, 440–453.

List of Authors

Maria Fe ANDRES
ICA, CSIC
Spain

Mireille ANSALDI
Laboratoire de Chimie Bactérienne
UMR 7283
CNRS Aix-Marseille University
France

Esther BADOSA
INTEA
University of Girona
Spain

Cédric BERTRAND
CRIOBE, UAR 3278
University of Perpignan
France

Anna BONATERRA
INTEA
University of Girona
Spain

Cécile CABASSON
BFP, UMR 1332
Bordeaux University
France

Pierluigi CABONI
Department of Life and
Environmental Sciences
University of Cagliari
Italy

Fernando CLAVIJO-COPPENS
GREENPHAGE SAS
France

Denis COSTECHAREYRE
GREENPHAGE SAS
France

Jean-Marc DEOGRATIAS
ASTREDHOR SUD-OUEST
France

Bernard DUMAS
LRSV
Toulouse University
France

Denis FAURE
I2BC
Paris-Saclay University
France

Hikmat GHOSSON
CRIOBE, UAR 3278
University of Perpignan
France

Azucena GONZÁLEZ COLOMA
ICA, CSIC
Spain

Nicolas GUIBERT
ASTREDHOR SUD-OUEST
France

Xavier LATOUR
CBSA UR4312
Normandy University
France

Hongwei LIU
Hawkesbury Institute for
the Environment
Western Sydney University
Australia

Patrice MARCHAND
ITAB
France

Léa MERLET
ITAB
France

Emilio MONTESINOS
INTEA
University of Girona
Spain

Laura MONTESINOS
INTEA
University of Girona
Spain

Nikoletta NTALLI
Department of Agriculture Crop
Production and Rural Environment
University of Thessaly
Greece

Pierre PÉTRIACQ
BFP, UMR 1332
Bordeaux University
France

Maria PLA
INTEA
University of Girona
Spain

Claire PRIGENT-COMBARET
EM, UMR 5557
Lyon 1 University
France

Diane ROBIN
ITAB
France

Maïder SAINT JEAN
UMR 6060
Bordeaux University
France

Marie-Virginie SALVIA
CRIOBE, UAR 3278
University of Perpignan
France

Ioannis A. STRINGLIS
Department of Biology
Utrecht University
Netherlands

Nicolas TAVEAU
GREENPHAGE SAS
France

Clara TORRES-BARCELO
INRAE-PACA
Unité de Pathologie Végétale
Avignon
France

Ke YU
Henan University
China

Wei ZHONG
Nanjing Agricultural University
China

Index

W, X

Printed and bound by CPI Group (UK) Ltd, Croydon, CR0 4YY

13/09/2026

14960997-0005